Ruminant Nitrogen Usage

Subcommittee on Nitrogen
Usage in Ruminants

Committee on Animal Nutrition

Board on Agriculture

National Research Council

NATIONAL ACADEMY PRESS
Washington, D.C. 1985

National Academy Press • 2101 Constitution Avenue, NW • Washington, DC 20418

NOTICE: The project that is the subject of this report was approved by the Governing Board of the National Research Council, whose members are drawn from the councils of the National Academy of Sciences, the National Academy of Engineering, and the Institute of Medicine. The members of the committee responsible for the report were chosen for their special competences and with regard for appropriate balance.

This report has been reviewed by a group other than the authors according to procedures approved by a Report Review Committee consisting of members of the National Academy of Sciences, the National Academy of Engineering, and the Institute of Medicine.

The National Research Council was established by the National Academy of Sciences in 1916 to associate the broad community of science and technology with the Academy's purposes of furthering knowledge and of advising the federal government. The Council operates in accordance with general policies determined by the Academy under the authority of its congressional charter of 1863, which establishes the Academy as a private, nonprofit, self-governing membership corporation. The Council has become the principal operating agency of both the National Academy of Sciences and the National Academy of Engineering in the conduct of their services to the government, the public, and the scientific and engineering communities. It is administered jointly by both Academies and the Institute of Medicine. The National Academy of Engineering and the Institute of Medicine were established in 1964 and 1970, respectively, under the charter of the National Academy of Sciences.

This study was supported by the Agricultural Research Service of the U.S. Department of Agriculture, by the Center for Veterinary Medicine, Food and Drug Administration of the U.S. Department of Health and Human Services, and by the American Feed Industry Association, Inc.

Library of Congress Cataloging in Publication Data

National Research Council (U.S.) Subcommittee on
   Nitrogen Usage in Ruminants.
   Ruminant nitrogen usage.

   Bibliography: p.
   Includes index.
   1. Nitrogen in animal nutrition.   2. Ruminants—
Feeding and feeds.   I. Title.
SF98.N5N37   1985   636.2′08′52   85-21682

ISBN 0-309-03597-X

Preface

The Subcommittee on Nitrogen Usage in Ruminants of the Committee on Animal Nutrition was instructed to address the increasingly apparent need to bring together the newer data and knowledge on ruminant nitrogen metabolism in such a way that a systematic, quantitative approach to the rationing of ruminants for nitrogen could be set forth.

The charge to the Subcommittee included the need to produce a report that could be used by producers, technical service people in the feed and related industries, as well as extension personnel, teachers, students, and research scientists. The document was to serve as a base, upon which future refinement could be built, in such a way that continual improvement would occur. Whereas many countries around the world have proposed nitrogen or protein systems for ruminants, this effort was to evaluate each of those and incorporate those biological principles that are common to all ruminants.

During the course of the deliberations of this subcommittee, invaluable help has been provided by the staff of the Board on Agriculture, especially Selma P. Baron and Philip Ross. This has been a most complex task, and without the encouragement and assistance of these people, the document contained herein would not have emerged.

We especially appreciate the efforts of Werner G. Bergen who reviewed this report for the Committee on Animal Nutrition and provided useful information for consideration by the authoring subcommittee. Also, the encouragement of Joseph P. Fontenot and Robert R. Oltjen has been greatly appreciated; the former also served as the reviewer for the Board on Agriculture. As a result, we feel that we have produced a document that meets the stated objectives of the original charge. The assistance of Zaira Batchelder, Alice Jones, and Elaine Wylie in the preparation of drafts and manuscripts is gratefully acknowledged.

Subcommittee on Nitrogen Usage in Ruminants

LEONARD S. BULL, *Chairman*
University of Vermont

WILLIAM CHALUPA
University of Pennsylvania

FREDRIC N. OWENS
Oklahoma State University

LARRY D. SATTER
USDA, Dairy Forage Research Center

CHARLES J. SNIFFEN
Cornell University

ALLEN H. TRENKLE
Iowa State University

DALE R. WALDO
USDA, Agricultural Research Service

Contents

Figures, Tables, and Appendix Tables

APPENDIX TABLES

Ruminant Nitrogen Usage

Introduction

The complexity of the digestive process in ruminants is well recognized by all who work with ration balancing in those species. The fact that digestion is a two step process—first by bacteria in the digestive tract and second by the host animal—results in the need to consider two entirely separate but interdependent ecosystems.

Nitrogen (protein) nutrition in ruminants is a complex, dynamic process. Extensive research has been and is being conducted on various components of the system.

Nitrogen (N) is a critical nutrient in the ruminant, since it is a key component in protein (amino acids). The ruminant cannot use nitrogen as a nutrient at the tissue level any more than the alfalfa plant can use atmospheric nitrogen without microbial intervention. However, as the nitrogen-fixing bacteria in the alfalfa roots enable atmospheric gaseous nitrogen to be trapped as plant tissue protein, the bacteria in the rumen can cause nonprotein nitrogen (primarily as ammonia) to be trapped as bacterial protein. These bacteria are subsequently digested by the animal and their protein is used to supply needed amino acids for production of animal protein for deposition in milk, wool, or animal or fetal tissues.

Ruminal bacteria in most instances cannot produce enough protein to meet the needs for maximum production of the animal. In such cases, productivity of the animal depends on the ability of the livestock producer to select those feeds and supplements to maximize bacterial production and, if needed, supply protein that will escape digestion in the rumen and pass to the small intestine to supply additional amino acids. If production of ammonia in the rumen from the feed sources, such as nonprotein nitrogen or feed protein that is rapidly digested, exceeds the capacity of the bacterial population to use ammonia, some of the excess can be lost and high concentrations may be toxic. A portion of bacterial nitrogen is in the form of nonprotein nitrogen and is of little nutritional value to the animal. Hence, conversion of dietary protein to microbial protein can be wasteful quantitatively and qualitatively.

The large intestine is another site of bacterial growth.

Unlike bacteria from the rumen, bacteria from the large intestine are excreted by the host without being exposed to the digestive processes in the small intestine, so fermentation in the large intestine increases recovery of energy but not nitrogen.

In the past, protein requirements for ruminants have been defined in terms of total or digestible nitrogen content of the feed. One unit of feed nitrogen is found in 6.25 units of feed protein. This system ignores the differences that exist among feedstuffs both in the form of nitrogen in feeds and its fate following ingestion by the animal.

This publication will review the biology of nitrogen metabolism in ruminants and outline a method for balancing the diets for ruminants based on these concepts. This method considers that the system is multicompartmental and dynamic. Critical variables will be identified, discussed, and averaged in an attempt to make the factors quantitative. Finally, guidelines will be proposed for formulating diets.

The derived system, although fundamentally logical, contains many constants. Variability in and interactions among these transfer coefficients are largely untested. Field application of the system must await further research with a wide variety of feedstuffs and ruminant classes. Consequently, this publication will attempt to describe the biological system, identify the limitations of our information, and propose a system of calculation that needs to be tested, modified, and improved in the future.

During these deliberations, several areas were discussed that are not considered here, even though we know that when data are available they will be factors. Such factors as environment and climate, stress, and a variety of other conditions that influence the endocrine balance and thus the metabolism of protein are examples. Also, no consideration was given to the many feed additives that may influence protein metabolism; these are discussed in other publications by the National Research Council.

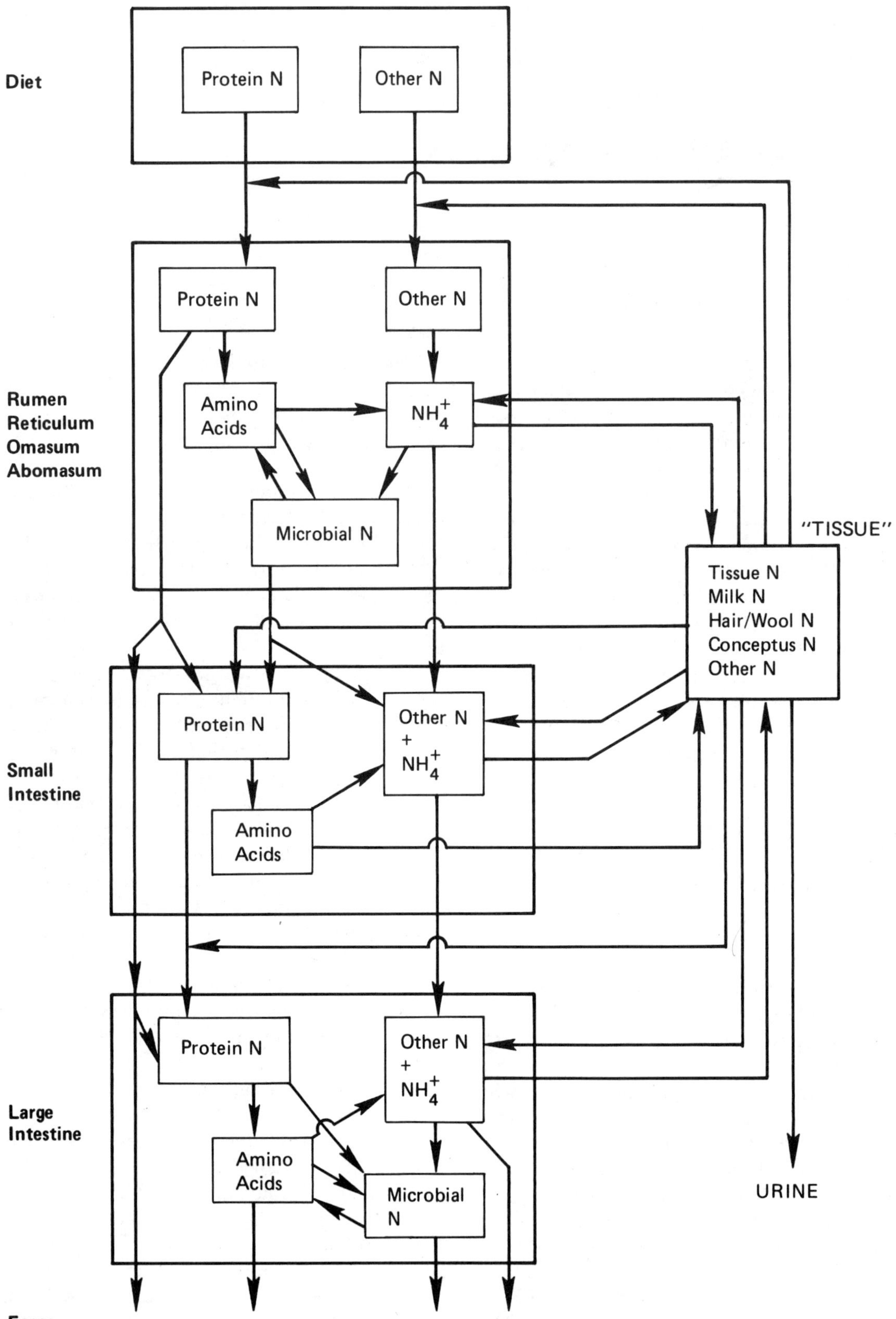

FIGURE 1   Schematic of nitrogen flow in the ruminant.

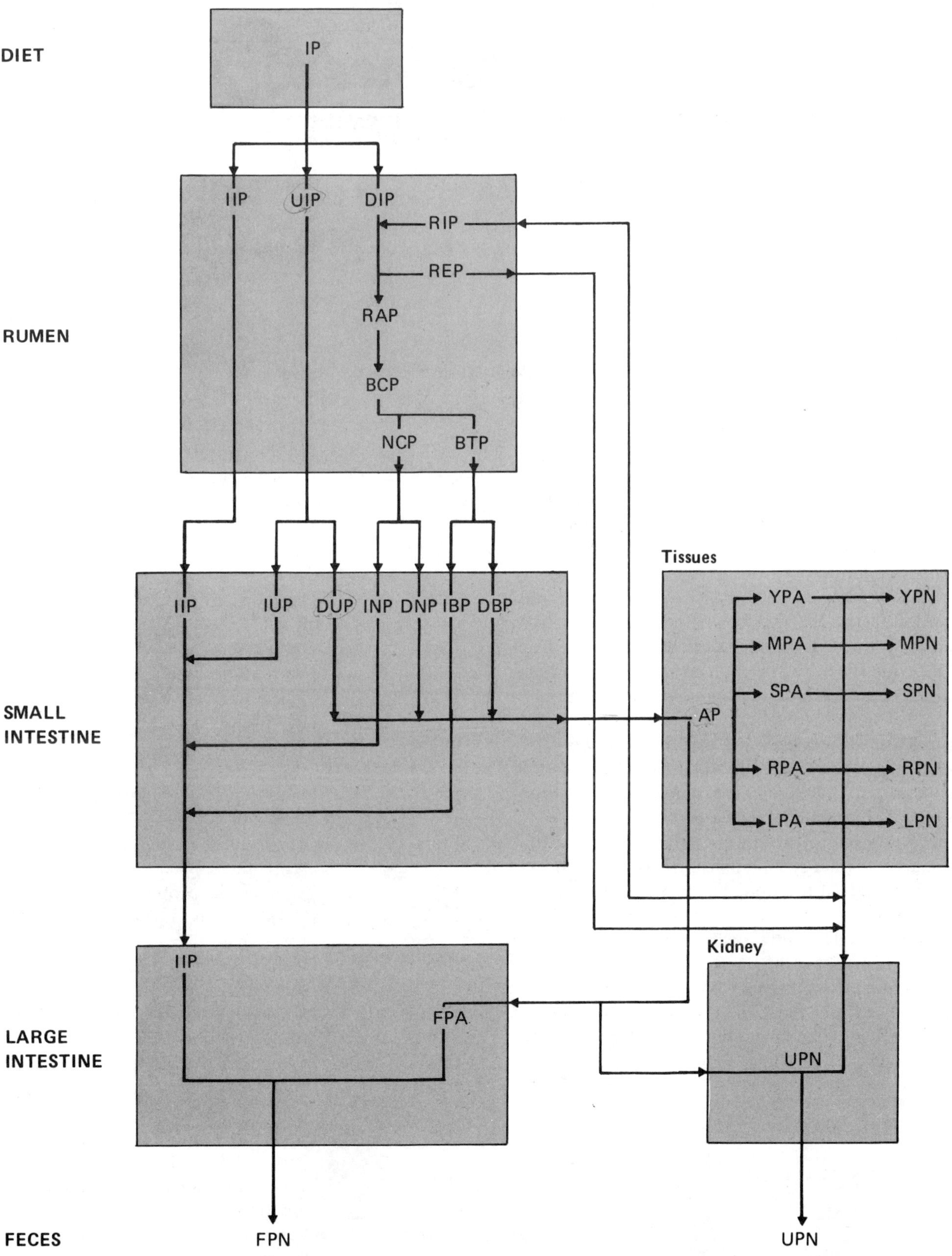

FIGURE 2   Schematic diagram of nitrogen flow in the ruminant using symbols developed in this publication.

Parameter Names for Describing Protein Metabolism

Parameter names for the fractions, or pools, and transfer coefficients that are compatible with computer use are suggested. A maximum of six characters is allowed in Fortran programming. Only letters and numbers are allowed, and a letter must be the first character; other characters in common use, such as parentheses and subscripts, are not acceptable. Terms for fractions, or pools, are limited to three characters and generally have units of mass/unit of time, such as grams or kilograms/day. Terms for transfer coefficients are limited to six characters with an implied slash separating the final fraction from the initial fraction and have proportional units. Fractions making up maintenance and production requirements are explicitly described as being in either units of absorbed or net protein. The word *absorbed* means the protein absorbed from the digestive tract. Absorbed is generally synonomous with metabolizable, but metabolizable was not used because its meaning relative to protein metabolism differs from that relative to energy metabolism. *Net* means the actual protein in that fraction. The word *crude* means N $\times$ 6.25. Names are intended to be consistent with the energy terms used in *Nutritional Energetics of Domestic Animals and Glossary of Energy Terms* (National Research Council [NRC], 1981). Many of the names currently used in either the Agricultural Research Council (ARC, 1980) or Proteines Digestibles dans l'Intestin (PDI) (Vérité et al., 1979) systems do not meet these requirements.

FRACTIONS OR POOLS

Term	Meaning
AP	Absorbed protein
ATDN	Adjusted total digestible nutrients (decreased 8 percent as relative units)

Term	Meaning
BCP	Bacterial (and protozoal) crude protein
BTDN	Baseline total digestible nutrients (1 $\times$ maintenance)
BTP	Bacterial (and protozoal) true protein
DBP	Digestible (true) bacterial (and protozoal) protein
DIP	Degraded intake (crude) protein (rumen)
DM	Dry matter
DMI	Dry matter intake
DNP	Digestible nucleic (acid crude) protein (intestine)
DOM	(Apparently) digested organic matter (total tract)
DUP	Digestible undegraded (crude) protein (intestine)
FOM	(Apparently) fermented organic matter (rumen)
FP	Fecal (crude) protein
FPA	(Metabolic) fecal protein (in) absorbed (protein units)
FPN	(Metabolic) fecal protein (in) net (protein units)
IBP	Indigestible (true) bacterial (and protozoal) protein
IDM	Indigestible dry matter (total tract)
IDMI	Indigestbile dry matter (total tract) intake
IIP	Indigestible intake (crude) protein (from ADIN or PIN analysis)
INP	Indigestible nucleic (acid crude) protein (intestine)
IOM	(Apparently) indigestible organic matter (total tract)
IOMI	(Apparently) indigestible organic matter (total tract) intake

Term	Meaning
IP	Intake (crude) protein
IUP	Indigestible undegraded (crude) protein (intestine)
LPA	Lactation protein (in) absorbed (protein units)
LPI	Lactation protein increment (LPA-LPN)
LPN	Lactation protein (in) net (protein units)
MPA	Maintenance protein (in) absorbed (protein units)
MPN	Maintenance protein (in) net (protein units)
NCP	Nucleic (acid) crude protein
NEL	Net energy (for) lactation
RAP	Ruminally available (nitrogen as) protein
REP	Rumen efflux (crude) protein (ammonia if positive, urea influx if negative)
RIP	Rumen influx (crude) protein (urea if positive, ammonia efflux if negative)
RP	Recycled (nitrogen as) protein
RPA	Retained protein (in) absorbed (protein units)
RPN	Retained protein (in) net (protein units)
SCP	Small (intestine) crude protein (BCP + UIP)
SPA	Surface protein (in) absorbed (protein units)
SPN	Surface protein (in) net (protein units)
STP	Small (intestine) true protein (BTP + UIP)
TDN	Total digestible nutrients
UP	Urinary (crude) protein
UIP	Undegraded intake (crude) protein
UPA	(Endogenous) urinary protein (in) absorbed (protein units)
UPN	(Endogenous) urinary protein (in) net (protein units)
YPA	Conceptus protein (in) absorbed (protein units)
YPN	Conceptus protein (in) net (protein units)

TRANSFER COEFFICIENTS

Term	Meaning
BCPDIP	Bacterial (and protozoal) crude protein/degraded intake (crude) protein
BCPDOM	Bacterial (and protozoal) crude protein/(apparently) digested organic matter
BCPFOM	Bacterial (and protozoal) crude protein/(apparently) fermented organic matter
BCPNEL	Bacterial (and protozoal) crude protein/net energy lactation
BCPRAP	Bacterial (and protozoal) crude protein/ruminally available (nitrogen as) protein
BCPTDN	Bacterial (and protozoal) crude protein/total digestible nutrients
BTPBCP	Bacterial (and protozoal) true protein/bacterial (and protozoal) crude protein
DBPBCP	Digestible bacterial (and protozoal true) protein/bacterial (and protozoal) crude protein
DBPBTP	Digestible bacterial (and protozoal true) protein/bacterial (and protozoal) true protein
DIPIP	Degraded intake (crude) protein/intake (crude) protein
DNPNCP	Digestible nucleic (acid crude) protein/nucleic (acid) crude protein
DOMDM	(Apparently) digested organic matter (total tract)/dry matter
DUPUIP	Digestible undegraded (crude) protein/undegraded intake (crude) protein
FOMDOM	(Apparently) fermented organic matter (rumen)/(apparently) digested organic matter (total tract)
FPADM	Fecal (metabolic) protein (in) absorbed (protein units)/dry matter
FPAIDM	Fecal (metabolic) protein (in) absorbed (protein units)/ indigestible dry matter
FPAIOM	Fecal (metabolic) protein (in) absorbed (protein units)/(apparently) indigestible organic matter (total tract)
FPIP	Fecal (crude) protein/intake (crude) protein
FPNDM	Fecal (metabolic) protein (in) net (protein units)/dry matter
FPNFPA	Fecal (metabolic) protein (in) net (protein units)/fecal (metabolic) protein (in) absorbed (protein units)
IIPIP	Indigestible intake (crude) protein/intake (crude) protein
IPDM	Intake (crude) protein/dry matter

Term	Meaning
LPNIP	Lactation protein (in) net (protein units)/intake (crude) protein
LPNLPA	Lactation protein (in) net (protein units)/lactation protein (in) absorbed (protein units)
MPNMPA	Maintenance protein (in) net (protein units)/maintenance protein (in) absorbed (protein units)
NCPBCP	Nucleic (acid) crude protein/bacterial (and protozoal) crude protein
RIPIP	Rumen influx (crude) protein/intake (crude) protein
RPNRPA	Retained protein (in) net (protein units)/retained protein (in) absorbed (protein units)

Term	Meaning
SPNSPA	Surface protein (in) net (protein units)/surface protein (in) absorbed (protein units)
UIPIP	Undegraded intake (crude) protein/intake (crude) protein
UPIP	Urinary (crude) protein/intake (crude) protein
UPNUPA	Urinary (endogenous) protein (in) net (protein units)/urinary (endogenous) protein (in) absorbed (protein units)
YPNYPA	Conceptus protein (in) net (protein units)/conceptus protein (in) absorbed (protein units)

Comparison of New Protein Systems for Ruminants

INTRODUCTION

Several new theoretical protein systems have been proposed that have potential application to feeding ruminants. These new systems require several additional concepts that the current National Research Council (NRC) systems, such as that for dairy cattle (1978), do not. Dietary intake crude protein (IP) is either degraded (DIP) in the rumen, with partial or total conversion to bacterial and protozoal crude protein (BCP), or passed from the rumen as undegraded intake protein (UIP). Microbial growth in the rumen requires either DIP, which may include either dietary nonprotein nitrogen (NPN), or a net ruminal influx of endogenous urea as crude protein (RIP) from either saliva or across the rumen wall. Production of BCP associated with microbial growth is related to energy fermented in the rumen and is expressed most commonly as a function of apparently fermented organic matter (FOM). Excess DIP increases the concentration of ruminal ammonia and increases the ruminal efflux of ammonia as crude protein (REP) by absorption and passage. Production of BCP represents both a protein requirement and a subsequent source of protein for the tissue needs of the cow. Efficiency of ruminal utilization of protein is 1.0 when DIP exactly meets the BCP need. When DIP equals BCP need, RIP must equal REP. The theoretical efficiency of tissue utilization of protein is maximum when BCP and UIP exactly meet the cow's tissue need. The theoretical efficiency of producing milk is maximum when both efficiencies in rumen and tissue utilization are maximum, i.e., neither DIP nor UIP is excessive.

The new concepts require that ruminal undegradability (UIPIP) must be specified in addition to a total tissue protein requirement. As higher milk production requires more total protein and available DIP exceeds that converted to BCP, more undegradable protein sources increase the efficiency of N use. A similar situation prevails in the rapidly growing animal.

Objectives of this paper are (1) to present a comparable tabulation of factors used in five U.S. and five European factorial systems that are static or partly static systems; (2) to calculate, as an example, minimal dietary protein and optimal UIPIP based on factors for a 600-kg cow producing from 10 to 40 kg of milk per day; and (3) to compare the expected flow of N into the small intestine and into the sinks of milk, urine, and feces. Papers previously published (Vérité et al., 1979; Waldo, 1979; Chalupa, 1980a; Vérité, 1980; Waldo and Glenn, 1982) have compared factors of some systems. Vérité et al. (1979) compared the protein concentration in dry matter (DM) required for milk production from 15 to 35 kg, and Geay (1980) compared protein required for growth based on several systems. Waldo and Glenn (1982) compared the distribution of dietary protein and N to milk, urine, and feces in five European systems.

FACTORS IN AVAILABILITY OF ABSORBED PROTEIN

Factors from 10 systems are compared in Tables 1 and 2. The current NRC dairy system (Swanson, 1977, 1982; NRC, 1978) is included as a reference. Four new U.S. systems have been proposed. These systems will be called Burroughs (Burroughs et al., 1971, 1974, 1975a, b; Trenkle, 1982), Satter (Roffler and Satter, 1975a,b; Satter and Roffler, 1975; Satter, 1982), Chalupa (1975b, 1980a), and Cornell (Fox et al., 1982; Van Soest et al., 1982). Two new European systems—the ARC system in Great Britain (Roy et al., 1977; ARC, 1980) and the PDI grele system in France (Vérité et al., 1979)—are official proposals within each country. Kaufmann (1977b, 1979) has proposed a system in Germany, and Landis (1979) has proposed a system in Switzerland. Haselbach (1980) and Schurch (1980) also presented discussions relative to the proposal of Landis. Danfaer (1979) has outlined many factors in a model of protein utilization from Denmark. Danfaer et al. (1980), Madsen et al. (1977), and Möller and Thomsen (1977) also presented data from Denmark that will be used for some values not specified in the model of Danfaer. Danfaer does not propose a system but gives some factors in protein utilization. The Cornell system introduces dynamic factors.

The new systems require specification of several new factors to describe availability of protein at the intestine. The division of IP into DIP and UIP fractions must be specified. The proportional production of BCP from DIP (BCPDIP) must be specified. Production of BCP must be related to dietary energy, which frequently is expressed as either FOM or apparently digested organic matter (DOM). The division of BCP into nucleic acid N as crude protein equivalent (NCP) and bacterial and protozoal true protein (BTP) must be specified. If theoretical urinary and fecal N excretion are to be calculated, the digestible nucleic acid N as crude protein equivalent (DNP) must be specified.

Intake Protein per Unit of Dry Matter

The required IP concentration in the dietary DM (IPDM) in the newer systems is still variable and directly related to milk production as in the NRC system based on total protein or crude protein (CP) (Table 1). However, in the newer systems, high milk production can be sustained with less IP if UIPIP also is increased, and dry cows in low production can be fed less IP if UIPIP is decreased.

Undegraded and Degraded Intake Protein per Unit of Intake Protein

All of the new systems consider the dietary IP to be divided into undegraded (UIP) and degraded (DIP) fractions, with the proportional division represented by UIPIP and DIPIP. This division is most generally considered continuously variable (Table 1). The Chalupa and ARC systems place all proteins into four classes having UIPIP of 0.20 ± 0.10, 0.40 ± 0.10, 0.60 ± 0.10, and > 0.70. The Cornell system defines an indigestible intake protein (IIP) fraction in IP. The Cornell system also further subdivides DIP into soluble and potentially degradable subfractions and includes a dynamic degradation of the potentially degradable subfraction.

The proper division of IP into UIP and DIP is the major new input required for these new systems to be effec-

tive in practice. The derivations or sources of these data are not always specified. Burroughs et al. (1975b) have an extensive tabulation of feedstuff degradabilities for their system. Satter is collecting *in vivo* data for common dairy feeds used in the north central United States. Chalupa (1980a) has accepted the ARC tabulation of feeds into four classes. The PDI system (Demarquilly et al., 1978) uses an extensive tabulation of solubility and *in vitro* fermentability. Vérité and Sauvant (1981) proposed equations for calculating digestible protein reaching the intestine from IP and protein of concentrates soluble in salt solution. The other systems propose neither a source of undegradability data nor an analytical method for obtaining the data.

Crude Bacterial Protein per Unit of Degraded Intake Protein

Six systems assume no loss or gain of protein in the production of BCP from DIP (BCPDIP), or BCPDIP = 1.00, but the ARC and Chalupa systems assume BCPDIP of dietary urea to be 0.80 (Table 1). The Satter system assumes an RIP equivalent to 12 percent of dietary IP and 90 percent utilization of ruminal ammonia, or bacterial and protozoal crude protein/ruminally available nitrogen as protein (BCPRAP); if degradability = 0.7, then the net influx is $0.12 - 0.7(1.0 - 0.9) = 0.05$ for BCPDIP = 1.05. The Danfaer model assumes BCPDIP = 0.90. The Cornell system proposes a range of BCPDIP from 0.5 to 0.9. Such low efficiencies increase implied protein requirement by increasing estimates of ammonia absorption and urinary excretion.

It seems unrealistic to assume that degradation of protein can be optimized for high milk production so that conversion of DIP to BCP fully attains 1.00, even though this is the goal of any ideal protein system. Water passage from the rumen will elute some ammonia that must be replaced. The dynamic model of Baldwin et al. (1977a) indicates that one-fourth of ammonia leaves the rumen by passage and three-fourths by absorption in a 40-kg sheep fed a 22.5 percent CP alfalfa hay at 37.9 g of DM/h. Kaufmann (1977a) found duodenal N (g/100 g of feed N) = $34.2 + 1032.7/\text{IP}$ (percent of dietary DM) in 45 observations on lactating dairy cows; this equation implies a gain of total N in the rumen below 15.7 percent dietary IP and a loss of total N above 15.7 percent. Hogan (1975) described protein reaching the intestines (g/g IP) = $0.33 + 0.18$ DOM intake with r = 0.96 using sheep; assuming that DOM = 0.67 DM, this equation implies a gain of protein in the rumen below 14.2 percent IP and a loss of protein above 14.2 percent IP. Oyaert and Bouckaert (1960) described the percentage of protein N intake absorbed in

TABLE 1   Assumed Factors in the Availability of Absorbed Protein to Ruminants

Factor	Symbol	NRC (1978)	Burroughs (1971, 1974, 1975a,b)	Satter (1982)	Chalupa, (1975b, 1980a)	Cornell (Fox et al., 1982; Van Soest et al., 1982)	ARC (1980)	Danfaer (1979)	PDI (Vérité et al., 1979)	Kaufmann (1977b, 1979)	Landis (1979)
					Proportional units unless specified otherwise in factor column						
Intake protein/dry matter	IPDM	V[a]	V	V	V	V	V	V	V	V	V
Indigestible intake protein/intake protein	IIPIP	ns[b]	ns	ns	ns	V[c]	ns	ns	ns	ns	ns
Undegraded intake protein/intake protein	UIPIP	ns	V	.34[d]	V[e]	D[f]	V[e]	V	V	.30[d]	V
Degraded intake protein/intake protein	DIPIP	ns	V	.66	V	V[g]	V	V	V	.70	V
Bacterial crude protein/degraded intake protein	BDPDIP	1.00[h]	1.00[h]	1.05	1.00[i]	.5–.9[j]	1.00	.90	1.00[h]	1.00[h]	1.00[h]
Bacterial crude protein/apparently fermented organic matter	BDPFOM	ns	.25	ns	.15	D[k]	.1875	.225	ns	.20	ns
Fermented organic matter/apparently digested organic matter	FOMDOM	ns	.52	ns	.65	D[l]	.65	.68	ns	.675	ns
Bacterial crude protein/digested organic matter	BCPDOM	ns	.13[m]	.132[n]	.0975	D[o]	.122	.153	.135	.135	.135
Digested organic matter/dry matter	DOMDM	V	V	V	V	V	V	V	V	V	V
Bacterial true protein/bacterial crude protein	BTPBCP	ns	.80[m]	.80	.80	.80	.80	.85	.80	.80[p]	ns
Digestible true bacterial protein/bacterial true protein	DBPBTP	ns	.80[m]	.80	.75	.80	.70	.76	.70	.90	ns
Digestible true bacterial protein/bacterial crude protein	DBPBCP	.75	.34[m]	.64	.60	.64	.56	.64	.56	.72	.68
Nucleic crude protein/bacterial crude protein	NCPBCP	ns	.20[m]	.20	.20	.20	.20	.15	.20	.20	ns
Digestible nucleic protein/nucleic crude protein	DNPNCP	ns	ns	ns	1.00	ns	ns	.85	ns	.425[p]	ns
Digestible undegraded protein/undegraded intake protein	DUPUIP	.75	.90	.87	.75	.90	.70	.87	.78[q]	.90	.85
Fecal metabolic protein	FPA or FPN[r]	FPN = .068IDMI[s]	FPA = .012DMI[t]	ns[u]	FPN = .030DMI[v]	ns[w]	ns[x]	FPA = .012DMI[y]	FPA = .057IOMI[z]	FPA = .026DMI[y]	FPA = .0177DMI[y]

[a] V = variable.

[b] ns = not specified and no assumption implied.

[c] Bound fraction that is undegradable and indigestible.

[d] Assumed variable but an average value is given.

[e] Variable but with proteins placed in four classes having degradabilities of .80 ± .10, .60 ± .10, .40 ± .10 and <.30.

[f] Dynamic degradation of potentially degradable fractions as functions of time.

[g] Further split into soluble and degradable subfractions.

[h] Implicitly assumed 1.00.

[i] Assumed 1.00 for DIP but .80 for urea.

[j] Variable as a result of system of feeding protein.

[k] Dynamic function of fluid flow rate from the rumen and true FOM.

[l] True FOM is a dynamic function of digestion lags, digestion rates, and passage rates of carbohydrate fractions.

[m] Commonly cited is .104, which is .13 × .8 or BTPDOM rather than BCPDOM; commonly cited digestibility is .80, which is OBPBTP rather than DBPBCP.

[n] Calculated assuming a 12.8% IP, 67 percent TDN diet with DIPIP = .66 and BCPDIP = 1.05 so (.128 × .66 × 1.05)/.67 = .132.

[o] Dynamic function incorporating both footnotes k and l.

[p] Kaufman (1977b) assumes .90 but .10 of this is assumed potentially digestible nucleic acid protein equivalent so it is excluded here. This exclusion reduces the digestion of the remaining BTP to .72. Kaufmann (1977b) assumes the .10 unit to have a digestibility of .85 so the full .20 units may be considered to have a digestibility of .425.

[q] Vérité et al. (1979) consider UIP to have a variable digestibility, but if their equation 4 is substituted into their equation 5, the digestibility becomes constant at .78, which is equal to the mean of their variable range from .60 to .95.

[r] FPA, fecal metabolic protein absorbed; FPN, fecal metabolic protein net.

[s] IDMI, indigestible dry matter intake.

[t] FPA subtracted from AP available from feed and not included in maintenance requirement; DMI, dry matter intake.

[u] Fecal metabolic protein not specified but included in maintenance requirement as a higher endogenous urinary protein.

[v] FPN specified and one-third is included in maintenance requirement as net protein.

[w] Not specified explicitly.

[x] Fecal metabolic protein not specified and neither subtracted from AP nor included in maintenance requirement but probably considered in lower true digestibility.

[y] FPA specified and included in total requirement as AP.

[z] Vérité et al. (1979) assume (equation 4) total fecal N = .143 insoluble N intake + .004 DOM + .0091 indigestible OM. The first factor is fully considered by a digestibility of UIP as described in footnote q. The second factor of .004 DOM is less than the indigestible fraction (.005 DOM), which arises from their assumption about indigestibility of CBP. The final factor of .0091 indigestible OM may be considered FPA or 57 g/kg of indigestible OM. IOMI, indigestible organic matter intake.

TABLE 2 Assumed Factors in the Requirement of Absorbed Protein by Ruminants for Maintenance, Milk Production, and Body Weight Change

Factor	Symbol	NRC (1978)	Burroughs (1971, 1974, 1975a,b)	Satter (1982)	Chalupa, (1975b, 1980a)	Cornell (Fox et al., 1982; Van Soest et al., 1982)	ARC (1980)	Danfaer (1979)	PDI (Vérité et al., 1979)	Kaufmann (1977b, 1979)	Landis (1979)
		Proportional units unless specified otherwise in factor column									
Maintenance	MPA or MPN[a]										
Fecal metabolic protein Equation	FPN	.068IDMI[b]	0[c]	0[d]	.010DMI[e]	0[f]	0[g]	0[h]	0[i]	0[h]	0[h]
Amount (g/600 kg)	FPN	144	0	0	64	0	0	0	0	0	0
Urinary endogenous protein	UPA or UPN										
Equation	—	UPN = $2.75\,W^{.5}$[j]	UPN = $.88\,W^{.734}$	UPA = $2.4\,W^{.75}$	UPN = $.938\,W^{.75}$	UPN = $.88\,W^{.735}$	—[k]	UPA = $.75\,W^{.73}$	0	ns	ns
Amount (g/600 kg)	—	UPN = 67	UPN = 96	UPA = 291	UPN = 114	UPN = 96	UPN = 60.625	UPA = 80	0	UPA = 110	ns
Surface protein	SPA or SPN										
Equation	—	SPN = $.2\,W^{.6}$	0	0	SPN = $.2\,W^{.6}$	0	SPN = $.1125\,W^{.75}$	0	SPA = $3.25\,W^{.75}$	0	ns
Amount (g/600 kg)	—	SPN = 9	0	0	SPN = 9	0	SPN = 13.75	0	SPA = 395	0	ns
Total net protein (g/600 kg)	MPN	220	96	ns	187	96	75[l]	ns	ns	ns	MPA =
Efficiency	MPNMPA	.60	.47	ns	.70	.60	.75	ns	ns	ns	$.9\,W^{.75m}$
Total absorbed protein (g/600 kg)	MPA	367	205	291	267	162	100	80	395	110	109
Production											
Lactation protein											
Net protein	LPN	$.019 + .4\,F$[n]	.028 − .034	.030	$.019 + .4\,F$[n]	—[o]	.03	.034	.033	.035	ns
Efficiency	LPNLPA	.70	.95	.60	.70	.70	.75	.56[p]	.67	.70	ns
Absorbed protein	LPA	.047	.030–.036	.050	.047		.04	.064	.050	.050	.05
Amount (g AP/30 kg)		1,414	990	1,500	1,414		1,200	1,920	1,500	1,500	1,500
Conceptus protein											
Equation	YPN	$1.136\,W^{.7}$	ns	ns	$1,136\,W^{.7}$	ns	—[q]	ns	ns	ns	ns
Amounts (g/600 kg)	YPN	100	ns	ns	100	ns	—	ns	ns	ns	ns
Efficiency	YPNYPA	.60	ns	ns	.70	ns	.75	ns	ns	ns	ns
Absorbed protein (g)	YPA	167	107–157	ns	143	ns	—	ns	205	ns	ns
Weight change in lactation											
Gain equation	NP	.225 G[r]	ns	.15 G	ns	ns	.150 G	.12 G	ns	ns	ns
Loss equation	NP	.144 G	ns	.15 G	ns	ns	.112 G	.12 G	ns	ns	ns
Efficiency	NPAP	.60	ns	.60	ns	ns	.75	ns	ns	ns	ns
Tissue protein in growth											
Equation as NP	RPN	.16–.19 G	.11–.15 G	ns	.11–.15 G	—[s]	—[t]	ns	.135–.186 G	ns	ns
Efficiency	RPNRPA	.45	.47	ns	.60	.60	.75	ns	.60	ns	ns

[a] MPA, maintenance protein absorbed, MPN, maintenance protein net.
[b] IDMI, indigestible dry matter intake.
[c] FPA subtracted from AP available from feed and not included in maintenance requirement.
[d] Fedal metabolic protein not specified but included in maintenance requirement as a higher endogenous urinary protein.
[e] FPN specified and one-third is included in maintenance requirement as net protein, DMI, dry matter intake.
[f] Not specified explicitly.
[g] Fecal metabolic protein not specified and neither subtracted from AP nor included in maintenance requirement but probably considered in lower true digestibility.
[h] FPA specified and included in total requirement as AP.
[i] Vérité et al. (1979) assume (equation 4) total fecal N = .143 insoluble N intake + .004 DOM + .0091 indigestible OM. The first factor is fully considered by a digestibility of UIP as described in footnote q of Table 1. The second factor of .004 DOM is less than the indigestible fraction (.005 DOM) which arises from their assumption about indigestibility of CBP. The final factor of .0091 indigestible OM may be considered FMP or 57 g/kg of indigestible OM.
[j] W, Weight in kilograms.
[k] UPN = 6.25 ($5.9206 \log_{10} W - 6.76$) for cattle and 6.25 (.02348 W + .54) for sheep.
[l] The value, 75 g of MPN is greater than 60.625 + 13.75 due to rounding differences but it is the commonly cited value for the 600 kg cow.
[m] Equation for calculating total MPA for maintenance from weight.
[n] Fractional protein = .019 + .4 F where F is fractional fat.
[o] Actual protein determined.
[p] Not specified but calculated from milk N/(milk + urine N − (urinary endogenous N + purine N + pyrimidine N + absorbed NH$_3$–N − influx urea–N from GI tract)).
[q] YPN as g/day = TP × $.03437e^{-.00262t}$ where $\log_{10}$ TP = $3.707 - 5.698e^{-.00262t}$ and t = number of days since conception.
[r] G is gain as kg/day.
[s] NP = (.235 − .00026W)G where W is live weight and G is gain as kg/day.
[t] TP = G ($168.07 - .16869\,W + .0001633\,W^2$)(1.12 − .1223 G) where protein gain is g/day, W is live weight, and G is gain as kg/day.

the rumen = $-20.8 + 1.62$ NH$_3$ − N (mg/100 ml of rumen fluid) with r = 0.91 for n = 10 diets fed to sheep; this equation implies a gain of protein in the rumen below 13 mg NH$_3$ − N/100 ml and a loss above 13 mg NH$_3$ − N/100 ml.

Crude Bacterial Protein per
Unit of Fermented Organic Matter

The most common expression of BCP to microbial energy fermentation is as a function of FOM (BCPFOM). Five systems assume proportional BCPFOM that range from 0.15 to 0.25 (Table 1). The Satter, PDI, and Landis systems relate BCP to DOM (BCPDOM) and specify no proportions for BCPFOM. The Cornell system is dynamic for this factor. Thomas (1973), in his summary, calculated a mean of 0.20 for 27 diets fed to sheep and cattle; Kaufmann (1977a) calculated a mean of 0.20 for 11 diets fed to dairy cows. Diet does affect this factor. McMeniman et al. (1976) calculated means of 0.1375 for diets with high concentrates and 0.1962 for diets with fresh or dried forages. Chamberlain and Thomas (1980a) present a range from 0.0625 to 0.1375 for diets consisting of only hay-crop silages.

Fermented Organic Matter per
Unit of Digested Organic Matter

When BCP is based on FOM, it is necessary to assume a proportional FOM per unit of DOM (FOMDOM). These proportions range from 0.65 to 0.68 except in the Burroughs system, for which 0.52 is used (Table 1). The Cornell system is dynamic for this factor.

Crude Microbial Protein per
Unit of Digestible Organic Matter

The eight fully static systems assume proportional BCP per unit of DOM (BCPDOM) that range from 0.0975 to 0.153 (Table 1); excluding the low 0.0975 for the Chalupa system and the high 0.153 for the Danfaer model narrows the range from 0.12 to 0.135.

Digested Organic Matter per Unit of Dry Matter

Because BCP is a function of DOM and IP usually is specified as a proportion of dietary DM, some specification of the proportional DOM per unit of DM (DOMDM) must be made. No system describes this relationship well. For subsequent calculations in the latter sections of this chapter when no value is specified, a value of 0.67 will be used, taken from the analyses of Tyrrell and Moe (1975) on the data of Wagner and

Loosli (1967). Diets containing from 25 to 75 percent concentrate and being consumed from 2.82 to 4.05 times maintenance had total digestible nutrients (TDN) from 66 to 68 percent. Increasing the percentage of concentrate increased intake, but digestibility depression offset the expected increase of digestibility.

True Bacterial Protein per
Unit of Crude Microbial Protein

All systems, except the Landis system, split the BCP into BTP and NCP components with the proportional division represented by BTPBCP and NCPBCP. All systems specify 0.80 for BTPBCP and 0.20 for NCPBCP except the Danfaer model, which specifies 0.85 and 0.15, respectively (Table 1).

Digestible Bacterial Protein per
Unit of True Bacterial Protein

All systems, except the Landis system, specify digestible bacterial true protein (DBP) per unit of BTP (DBPBTP). These proportions range from 0.70 to 0.90 with 0.80 most frequently used (Table 1).

Digestible Bacterial Protein per
Unit of Crude Bacterial Protein

All new systems specify a proportional DBP per unit of BCP (DBPBCP) with the range from 0.56 to 0.72 (Table 1). All of the new proportions are below the 0.75 that is assumed in the present NRC system.

Digestible Nucleic Acid Nitrogen per
Unit of Crude Nucleic Acid Nitrogen

Only three systems specify any rate for NCP. The Danfaer model assumes a proportional 0.85 for digestible nucleic acid N as protein equivalent (DNP) per unit of NCP (DNPNCP). The Chalupa system assumes DNPNCP = 1.00. Kaufmann (1977b) describes his system as having a NCPBCP of 0.10 that has a digestibility of 0.85 and another NCPBCP of 0.10 that has a digestibility of 0. For a greater equivalence to the Danfaer model, the Kaufmann system is redescribed in Table 1 as having the full NCPBCP of 0.20 that has a digestibility of 0.425. This description gives the same distribution of nucleic acid N as the original Kaufmann system. Specification of the digestibility of nucleic acid N is necessary to compare the theoretical urinary and fecal excretion of N with *in vivo* results. It is also unreasonable to expect that the nucleic acid N will be more digestible than the bacteria that contain it.

*Digestible Undegraded Protein per
Unit of Undegraded Intake Protein*

All systems specify proportional digestible undegraded intake protein (DUP) per unit of UIP (DUPUIP) with the range from 0.70 to 0.90 (Table 1). The PDI system assumes variable digestibility as specified in their equation 5 (Verite et al., 1979). Substitution of their equation 4 into their equation 5 gives a true digestibility of undegraded N = [(0.65 − 0.143) × insoluble N intake]/0.65 × insoluble N intake = 0.78 for a mean; this mean is also equal to the mean of their variable range from 0.60 to 0.95. The difference between this mean and their variable digestibilities is that their variable digestibilities include all of the residual error of a particular feed associated with the derivation of the constants describing fecal N output.

Fecal Metabolic Protein

Fecal metabolic protein is the most variable factor in the systems (Table 1). Fecal metabolic protein either is not specified as a separate factor or is specified as a separate factor that is a function of DM intake (DMI), indigestible dry matter intake (IDMI), or indigestible organic matter intake (IOMI). The units are either as absorbed (FPA) or net (FPN) protein. The ARC and Satter systems do not specify any FPA or FPN as a separate factor. The Cornell system mentions it as a separate factor but does not specify an equation. The PDI system specifies FPA as 0.057 per unit of IOMI. The NRC system specifies FPA as 0.068 per unit of IDMI or as 0.030 per unit of DMI with slightly less accuracy. The Chalupa system specifies FPN as 0.03 per unit of DMI. However, it considers two-thirds of this to be undigested bacterial cells that are accounted for as indigestible bacterial protein and indigestible nucleic acid protein equivalent in this publication. All other systems specify FPA as a function of DMI with factors ranging from 0.012 to 0.026. The Burroughs system specifies the FPA as a reduction of absorbed protein (AP) from feed. All systems specifying FPA may use that FPA as a reduction of AP from feed.

In vivo data on lactating cows fed forage-concentrate diets may be used for comparing with the fecal metabolic protein specifications of these systems. Boekholt (1976) found that digestible protein (DP) as a percentage of dietary dry matter or DP (percent of DM) = $0.833 \times$ IP (percent of DM) − 3.31, with $r^2 = 0.95$, $s_{y \cdot x} = 0.469$, and n = 362 for dairy cattle whose mean milk production was 18.9 ± 5.2 kg/day (as a standard deviation) and IP (percent of DM) was 16 ± 2.3 percent. This equation implies a fecal metabolic protein fraction of 0.033 g/g of dietary DM. Waldo and Glenn (1982) per-

formed a similar analysis on the data of Conrad et al. (1960) and found DP (percent of DM) = $0.861 \times$ IP (percent of DM) − 2.86, with $r^2 = 0.92$, $s_{y \cdot x} = 0.743$, and n = 177 for dairy cattle whose milk production was 11.8 ± 3.1 kg/day and IP (percent of DM) was 15.3 ± 2.8. This equation implies a fecal metabolic protein of 0.029 g/g of dietary DM. That these proportions could round to 0.030 is support for the slightly less accurate estimate of the NRC system, but as FPA rather than FPN. At 15 percent dietary IP, these equations imply that 57 percent of the fecal N arises from this source and that 19 to 22 percent of dietary IP is required to meet this need. It seems realistic to include a factor for fecal metabolic protein, since it has so much quantitative importance.

FACTORS IN THE REQUIREMENT FOR ABSORBED PROTEIN

No factors other than those already used in the NRC system are required in the new protein systems. The total protein requirement includes that for fecal metabolic protein, maintenance, and production. The maintenance requirement may include fecal metabolic protein, urinary endogenous protein, and surface protein. The production requirement is the sum of one or more of four factors—lactation, conceptus, weight change, and growth (including surface material).

Maintenance

FECAL METABOLIC PROTEIN

Fecal metabolic protein is considered differently in the systems (Table 2). The ARC and Satter systems do not specify any FPA or FPN as a separate factor. The Cornell system does not specify how the fecal metabolic protein fraction is considered. The Burroughs system has specified FPA as a reduction from AP available from feed. The Kaufmann and Landis systems specify FPA as a component of total requirement independent of maintenance; the model of Danfaer also seems to include FPA as a component of total requirements. The inclusion of FPA either as a feed reduction factor or fully considered as a component of total requirement supplies the full requirement for fecal N. The NRC system includes a part of the fecal metabolic protein in the maintenance requirement and the remainder in the production requirement. The Chalupa system specifies one-third of total fecal metabolic protein in maintenance. The PDI system includes only part of total fecal metabolic protein as a component of the requirement; the remaining requirement for fecal N excretion must be met by reducing assumed urinary N excretion.

URINARY ENDOGENOUS PROTEIN

Neither the PDI nor Landis systems specify this factor (Table 2). All other systems specify an endogenous urinary protein equivalent either in units of absorbed (UPA) or net (UPN) protein. This requirement is usually calculated as a power function of body weight near 0.75.

SURFACE PROTEIN

No specification is made in the Landis system and a zero specification is made in the Burroughs, Satter, Cornell, Danfaer, and Kaufmann systems (Table 2). All other systems specify a surface protein requirement in units of absorbed (SPA) or net (SPN) protein. The SPN is less than 5 percent of the total maintenance for the Chalupa and NRC systems. The SPN represents about 20 percent of the maintenance requirement of the ARC system. Because the PDI system uses the SPA required to give N retention equal to hair and scurf loss as its total maintenance requirement, its SPA is relatively higher than all others.

TOTAL

The total maintenance requirement as AP is in Table 2 for comparative purposes. The total maintenance requirements are extremely variable, ranging from 100 to 395 g of AP. Unfortunately, these are not equivalent because some include fecal metabolic protein and some do not. The relatively low requirement of the Burroughs system can be accounted for partially by the deduction of fecal metabolic protein from available AP. Similarly, the relative low requirements of the Danfaer, Kaufmann, and Landis systems are accounted for by their consideration of fecal metabolic protein as a separate component of total requirement. No equivalent factors can account for the lowest requirement in the ARC system; however, the failure to include a fecal metabolic protein factor probably contributes to its smallness.

Production

LACTATION

The lactation protein requirement as absorbed (LPA) units is the assumed protein concentration or lactation net protein requirement as net (LPN) units divided by the assumed efficiency (LPNLPA) (Table 2). The most commonly assumed efficiency is 0.70. The efficiency of Burroughs is highest at 0.95 and of Danfaer is lowest at 0.56 (Table 2). The LPA requirement for 30 kg of milk is in Table 2. These requirements range from 990 to 1,920 g of LPA, with the major cause of differences being differences in efficiency. The ARC system has the second lowest requirement for milk along with the lowest requirement for maintenance and no reduction of available AP by fecal metabolic protein.

CONCEPTUS

The conceptus protein requirement as absorbed (YPA) units for the last 2 months of pregnancy varies from 107 to 205 g (Table 2). A frequent requirement is about 160 g. Five systems have not described a requirement for the conceptus.

WEIGHT CHANGES IN LACTATION

Six systems do not specify a factor for weight change. When weight change is specified as retained protein in net (RPN) units in the NRC, ARC, Danfaer, and Satter systems, the proportions range from 0.112 to 0.225. The validity of the NRC and ARC systems assuming a different proportion for gain and loss when the units are defined as RPN seems questionable. The efficiency for weight change is the same as for lactation in the ARC and Satter systems.

GROWTH

Six systems have proposed the proportional gain as protein (Table 2). Burroughs et al. (1974) assume the proportional retained protein as net (RPN) units in liveweight gain (G) declined from 0.150 at 150 kg of liveweight to 0.110 at 500 kg for finishing steers and heifers of early maturing breeds. Chalupa (1975b) adopted these same data. The ARC (1980) assume proportional RPN in empty body weight gain (EBWG) declined from 0.181 at 50 kg of empty body weight (EBW) to 0.140 at 500 kg for steers of an average size with 0.6 kg EBWG/day. Proportional RPN is changed by a factor of 0.90 for smaller breeds and 1.10 for larger breeds. Proportional RPN is changed by a second factor of 0.90 for heifers and 1.10 for bulls. Proportional RPN is changed by a third factor of 0.013 subtracted from 1.0 for each 0.1 kg of EBWG greater than 0.6 and 0.013 added to 1.0 for each 0.1 kg of EBWG less than 0.6. The PDI (Vérité et al., 1979) system assumes proportional RPN in G to decrease from 0.186 to 0.135 with maturity; the proportion varies with liveweight, G, breed, and sex. Robelin and Daenicke (1980) extend the PDI system by giving a set of equations for describing the proportional RPN as continuous functions of liveweight and G within very early maturing steers, early maturing bulls, and late maturing bulls. Fox et al. (1982) describe the retention of RPN as a function of EBW for steers of medium frame size.

This steer of medium-frame size is considered a reference animal with an equivalent weight equal to its actual weight. Eight other frame sizes and two other sexes are specified that require adjustment factors for converting their actual weight to equivalent weight; these equivalent weights, theoretically, have the same body composition. Adjustment factors for steers are 1.25 for smallest frame, 1.00 for medium frame, and 0.83 for largest frame. Adjustment factors for heifers are 1.56 for smallest frame, 1.25 for medium frame, and 1.04 for largest frame. Adjustment factors for bulls are 1.04 for smallest frame, 0.83 for medium frame, and 0.69 for largest frame. The NRC (1978) requirement for dairy cattle is in Table 2. The NRC (1984) requirement for beef cattle calculates RPN as a function of the energy concentration of gain for steers. Composition of gain of medium-frame heifers is assumed equivalent to medium-frame steers weighing 15 percent more. Composition of gain of bulls and large-frame steers was assumed equivalent to medium-frame steers weighing 15 percent less. Liveweight, daily gain, breed or frame size, and sex are the four most important factors affecting the protein energy ratio in growing and fattening cattle. The functional change in protein and fat depositions with increasing energy deposition remains somewhat controversial. The proposals range from linear changes in energy deposited as fat and protein (Tyrrell et al., 1974; Geay, 1984) to an asymptotic maximal deposition of protein (Byers, 1982b) and to a maximal protein deposition followed by a decrease (Anrique, 1976).

The efficiencies of converting retained protein as absorbed (RPA) units to RPN, or RPNRPA, range from 0.45 (NRC, 1978) to 0.75 (ARC, 1980) in Table 2. The NRC (1984) requirements for beef cattle assume an efficiency of 0.66.

Data on sheep are not specific in the various models. As a consequence they are not covered in this discussion. Comments on gain and concepts applying to it would be appropriate for sheep in lieu of more definite data.

DYNAMIC MODELS

Dynamic models have been proposed that describe protein utilization for the entire animal. Other dynamic models describe ruminant digestion of dietary crude protein and carbohydrates, while others describe nitrogen metabolism in the ruminant without any reference to energy. Some of these models are considered preliminary. Generally, the models are not published in full detail so that direct communication with the authors is required for enough detail to use, compare, or challenge them.

Two dynamic models for specifying protein requirements for ruminants are being developed in the United States. At Michigan, Fox et al. (1976) introduced a net protein system. Bergen et al. (1979) calculated the upper limit of ruminal microbial protein synthesis. Bergen et al. (1982) describe the efficiency of microbial protein synthesis in relation to specific growth rate, growth yield, and maintenance in rumen bacteria. They demonstrate an increase in ribonucleic acid/protein ratio as specific growth rate increases in anaerobic bacteria. Such a large difference in this ratio must raise questions about the constancy of the ratio of nonammonia nitrogen and amino nitrogen entering the small intestine or apparently absorbed in the small intestine. Johnson and Bergen (1982) describe the effects of diet on the fraction of organic matter digestion occurring in the rumen and efficiency of microbial protein production. Waller et al. (1982) describe their progress toward a dynamic model of protein requirement for the ruminant that considers economics as well as nutrition and emphasizes the algebra and linear programming necessary to consider the uncertainty of feed composition and least cost formulation.

At Cornell, Van Soest et al. (1982) propose a rumen submodel for nitrogen utilization that describes the output of protein by using the following inputs: soluble protein; three true protein subfractions based on the degradability rates (B_1, rapid; B_2, intermediate; and B_3, slow); bound protein; nonstructural carbohydrate; potentially digestible organic matter; rates of digestion for each protein, nonstructural carbohydrate and potentially digestible organic matter subfractions; and rates of passage for liquids and solids. Fox et al. (1982) complete the total model by describing the factors in the calculation of requirements for growth (Table 2).

The nitrogen flux within the rumen as REP loss of ammonia by absorption and passage and as RIP gain of urea from saliva or blood are very important. Nolan et al. (1976) describe the nitrogen dynamics on a three pool model of rumen ammonia, plasma urea, and cecal ammonia in a sheep eating about 22 g air dry feed/h that contained 18.7 percent CP. When mean dietary N intake was 16.3 g/d, mean rumen ammonia N was 20.9 mg/100 ml, mean plasma urea N was 18.1 mg/100 ml, and total flux of rumen ammonia was 15.0 g/d. Of this total flux 28.7 percent was recycled, and 71.3 percent was irreversible loss via influx and efflux. The influx sources, as a percentage of total, were: dietary and endogenous sources, 61.9; blood urea, 6.9; and from cecal ammonia but not via blood urea, 2.5. The efflux losses, as a percentage of total, were: absorbed, 44.4; microbial protein synthesized into tissue, 20.6; and cecal ammonia from rumen microbes, 6.3. Only 40 percent of rumen bacterial N came from ammonia N, and only 20 percent of urea degraded in the intestinal tract was degraded in the rumen. Mazanov and Nolan (1976) describe the nitrogen dynamics in a nine-part model using the above data combined with data from lower N intakes. When mean dietary N intake was 14.16 g/d, total flux of ammonia N was 9.11 g/d. Of this total flux, 19.2 percent was recycled. The influx sources, as a percentage of total, were: dietary amino N, 67.0; urea, 13.2; and dietary ammonia N, 0.6. The efflux losses, as a percentage of total, were: microbial N not recycled, 31.3; and ammonia N absorbed or passed, 49.5.

Baldwin et al. (1977a) proposed a model of ruminant digestion that uses 12 chemical inputs: lignin, cellulose, hemicellulose, pectin, starch, soluble carbohydrate, organic acids, lipids, ash, insoluble protein, soluble protein, and NPN. The model uses one physical input: fraction retained on 1-mm sieve. This model emphasizes the importance of ammonia passage as a loss of N from the rumen. Baldwin and Denham (1979) present another model of N metabolism in the rumen that emphasizes the difference in affinity of the two major enzymes for ammonia. Glutamic dehydrogenase is constitutive and has a low affinity for ammonia (K_m = 5 mM); glutamine synthetase is induced at low ammonia concentrations and has a high affinity for ammonia (K_m = 0.2 mM). Such a difference may explain why microbial growth is not limited until concentrations fall below 3 to 5 mg/100 ml, but microbial fermentation of the carbohydrates in some diets is limited at concentrations below 20 to 25 mg/100 ml, a critical point in comparing *in vitro* and *in vivo* results. The dietary differences in ruminal methylamine concentration (Hill and Mangan, 1964) may affect the competitive uptake of ruminal ammonia by bacteria.

Black et al. (1980–1981) describe a model of rumen function that uses these chemical inputs: beta-hexose (lignin, cellulose, and hemicellulose); alpha-hexose (pectin and starch); soluble carbohydrate (including glycerol); total fatty acids; inorganic sulfur; ash; protein (true protein and free amino acids); NPN (including nucleic acids); potential degradability of beta-hexose; and potential degradability of protein. The model has one physical input: modulus of fineness of diet. Other modeling inputs are: feed intake; time feeding; time ruminating; and reduction in maximum rate or degradation of beta-hexose, alpha-hexose, and protein due to diet. Endogenous inputs are: true protein, NPN, and inorganic sulfur. Beever et al. (1980, 1981) did a sensitivity analysis for 22 variables that could not be set with confidence. Six variables with a high sensitivity, i.e., a change in protein flow greater than 40 percent from the possible range in input variable, were: potential degradability of protein, fractional outflow rate of water, fractional outflow rate of microbes, energy required for microbial maintenance, salivary flow, and proportion of rumen ammonia available for microbial growth.

Faichney et al. (1980) found predictions of this model to be closer to observations in one data set than predictions from the ARC and PDI systems.

COMPARISON AND CHALLENGE OF SYSTEMS WITH *IN VIVO* DATA

A comparison and challenge of implications of the systems with *in vivo* data from lactating cattle is informative after their assumptions and calculations are understood. For these comparisons and challenges we have calculated IP as a percentage of DM, optimum UIPIP, either the sum of BTP and UIP or the sum of BCP and UIP reaching the small intestine, fecal N as a percentage of dietary N, urinary N as a percentage of dietary N, and milk N as a percentage of dietary N for a 600-kg cow producing 10, 20, 30, and 40 kg milk/day with degradability optimal. These predicted data then are compared with expected *in vivo* data on protein reaching the small intestine (Tamminga and van Hellemond, 1977; Journet and Vérité, 1979; Rohr et al., 1979) and *in vivo* data on the distribution of N in feces, urine, and milk (Conrad et al., 1960; Boekholt, 1976). Some additional data and assumptions are required because BCP production is a function of FOM and fecal metabolic protein is a function of DMI, IDMI, or IOMI. No attempt was made to compare and challenge the Cornell system because it contains several dynamic relationships.

Energy Standards

Except for the ARC and PDI systems, the new protein systems are published without any specific statement of or reference to an energy standard. Production of BCP is related to energy fermented in the rumen, which is more frequently FOM. Fecal metabolic protein is related to some dietary component, most frequently dietary DM. These or other required energy variables must be specified for a complete system. The energy requirements and the DM intakes used in these comparisons and challenges of protein feeding systems and their sources are in Table 3. Chalupa (1980a) used metabolizable energy as the energy unit, but energy requirements were not fully elaborated, so TDN is used as the energy requirement for the Burroughs, Chalupa, NRC, and Satter systems. Assumptions about concentrations of energy in dietary DM vary as well as the assumptions about absolute amounts of either. In going from 5 to 40 kg of milk, energy concentrations increase 85 percent in the PDI system, 33 percent in the Kaufmann system, and 65 percent in the Landis system. When energy concentration is not specified, we have assumed it to be constant (based on the proposed maximum of Tyrrell and Moe, 1975) with DOMDM = 0.67 as discussed earlier for TDN.

TABLE 3   Dry Matter Intakes and Energy Standards When Energy Concentration Varied as Used in Comparison and Challenge of Protein Systems

Milk (kg)	NRC[a] Dry Matter (kg)	ARC[b] Dry Matter (kg)	Danfaer[c] Dry Matter (kg)	PDI[d] UFL	PDI[d] Dry Matter (kg)	Kaufmann[e] SE	Kaufmann[e] Dry Matter (kg)	Landis[f] NEL (MJ)	Landis[f] Dry Matter (kg)
5	8.6	7.2	7.5	7.1	12.3	4375	9	51.2	11.5
10	10.9	9.4	9.8	9.3	13.7	5750	11	66.9	13.0
15	13.2	11.7	12.1	11.5	15.1	7125	13	82.6	14.5
20	15.4	14.1	14.4	13.7	16.5	8500	15	98.3	16.0
25	17.7	16.4	16.7	16.1	17.9	9875	17	114.0	17.5
30	20.0	18.8	19.0	18.6	19.2	11250	19	129.7	19.0
35	22.2	21.3	21.3	21.0	20.6	12625	20	145.4	20.5
40	24.5	23.6	23.6	23.5	21.9	14000	21	161.1	22.0

[a] Calculated from total digestible nutrients for maintenance of the mature, lactating, 600-kg cow and production of milk with 3.5 percent fat (NRC, 1978) by dividing by .67. Used for the NRC, Burroughs, Chalupa, and Satter systems.

[b] Calculated from megajoules of metabolizable energy for maintenance of a 600-kg cow with 0 live-weight change and production of milk with 3.68 percent fat while being fed a diet with metabolizability or q = .60 (ARC, 1980) by dividing by 11.

[c] Calculated as Scandinavian feed energy (FE) units from Madsen et al. (1977) (Table 3) starting from N in microbial net protein divided by 16 g N per FE divided by digestibility or .60 divided by efficiency or .71; then 16.5 FE = 19 kg dry matter from Danfaer et al. (1980, p. 12).

[d] From Vérité et al. (1978, Table 12.3). UFL = the French net energy unit and is the total requirement for a 600-kg cow consuming good quality forage and producing milk with 4.0 percent fat.

[e] From Kaufmann (1977b, 1979). SE = starch equivalent unit. Data are linearly interpolated and extrapolated from the data in Table 2 (Kaufmann, 1979).

[f] From Landis (1979). NEL = net energy for lactation. Data are linearly interpolated and extrapolated from data in Table 2.

Such different energy assumptions contribute to differences among the protein systems.

While the use of TDN is questioned by many, the available data for alternatives are not as numerous. Of even more importance is the fact that in use many of the alternative energy terms are derived from TDN or an estimate of TDN. Thus, we do not feel that TDN is, in fact, an improper base.

Additional Assumptions

Several additional assumptions that were required in one or more of the systems and their bases are in Table 4. Where the disposition of NCP is not described, its digestibility was assumed to be 0.85, and the excretion of digested fraction was assumed to be via urine.

Minimum Dietary Intake Protein Percentage

Dietary IP percentages required in nine systems, when undegradability is optimal, are compared (Figure 3). Differences among the systems are smaller (from 9 to 13.2 percent IP) at 10 kg of milk but become larger (11 to 17.4 percent IP) at 40 kg of milk. The Danfaer model requires the highest IP percentage at every milk yield.

Presumably, this higher requirement is primarily a result of a lower assumed efficiency of milk production. The Burroughs, ARC, and Satter systems require a much lower IP percentage than other systems at higher milk production. The probable causes of their low requirements are the highest efficiency for converting AP to milk assumed in the Burroughs system; the second lowest AP requirement for lactation plus a low AP requirement for maintenance with no separate fecal metabolic protein requirement in the ARC system; and no fecal metabolic protein requirement as a function of DM intake either alone or as a component of maintenance in the Satter system.

Optimum Undegradability

Optimum IP undegradabilities required when IP percentage is minimum are compared (Figure 4). Differences among the systems are large (7 to 41 percent at 10 kg of milk) and remain large (20 to 55 percent at 40 kg of milk). The undegradability of many common diets for dairy cows is considered to be near 0.30 (Satter and Roffler, 1975). The ARC and Burroughs systems do not require an undegradability as high as 0.30 at 40 kg of milk per day. These low undegradability requirements are

TABLE 4 Additional Assumptions of Protein and Energy Relationships

Assumption	Systems[a]
Protein[b]	
CBPDIP = 1.00	N
DNPCNP = .85[c]	A, P, B, S
LNPLMP = .60	L
Energy[d]	
DM = TDN/.67[e]	N, C, S
11 MJ ME/kg DM[f]	A
19 kg DM = 16.5 FE[g]	D
OM = DM × .9	P
DE = ME/.82[h]	A
DOM = DM × .67	D
DOM = UFL × .732[i]	P
DOM = NEL/9.31[j]	L
1 kg DOM = 900 SE[k]	K
19 MJ ME/kg DOM[l]	A
IOM = OM − DOM	P

[a] A, ARC; B, Burroughs; C, Chalupa; D, Danfaer; K, Kaufmann; L, Landis; N, NRC; P, PDI; and S, Satter.

[b] CBPDIP, crude bacterial protein/degraded intake protein; DNPCNP, digestible nucleic acid bacterial protein/crude nucleic acid bacterial protein; LNPLMP, lactation net protein/lactation metabolizable protein.

[c] From Danfaer (1979).

[d] DM, dry matter; TDN, total digestible nutrients; ME, metabolizable energy; FE, Scandinavian feed energy unit; OM, organic matter; DE, digestible energy; DOM, apparently digested organic matter; UFL, French net energy unit; NEL, Swiss net energy unit; SE, starch equivalent; IOM, indigestible organic matter.

[e] From analyses of Tyrrell and Moe (1975) on data of Wagner and Loosli (1967).

[f] From ARC (1980, see p. 112).

[g] From Danfaer et al. (1080, see p. 12).

[h] From ARC (1980, see p. 136).

[i] From INRA (1978, see p. 589); and ARC (1980, see Table 4.7). ME = 2.73 UFL and DOM = ME/3.72 so DOM = UFL × (2.73/3.72) = UFL × .732.

[j] Calculated from Landis (1979). .135 CBPDOM/.0145 CBPNEL = 9.31.

[k] From Kaufmann (1977b).

[l] From ARC (1980, see p. 136).

related to the low protein percentages, and their causes as discussed earlier. The Chalupa system always requires an undegradability greater than 0.30. This high undegradability results primarily from the low BCPDOM.

A plot of the UIPIP as a function of concentration of IP (Figure 5) indicates a great diversity among the systems. The differences are largely caused by assumptions about changes of energy concentration and dry matter intake for meeting the additional energy needs for high milk production. If increasing energy requirements are met by increasing energy concentration more than DM intake, as in the Kaufmann, Landis, and PDI systems, then protein concentration varies more than UIPIP. If

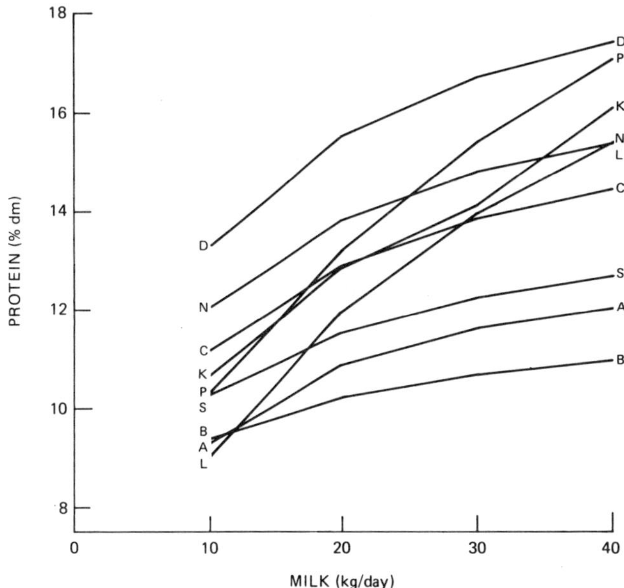

FIGURE 3 Intake protein percentage in dry matter as a function of milk production. A, ARC; B, Burroughs; C, Chalupa; D, Danfaer; K, Kaufmann; L, Landis; N, NRC; P, PDI; and S, Satter.

increasing energy requirements are met by increasing DM intake more than energy concentration, as in the ARC, Burroughs, Chalupa, Danfaer, and Satter systems, then UIPIP varies more than protein concentration.

Protein Reaching the Small Intestine

IN VIVO REFERENCE DATA

Three data sets (Tamminga and van Hellemond, 1977; Journet and Vérité, 1979; Rohr et al., 1979) are available that describe protein flowing into the duodenum of the lactating cow. Tamminga and van Hellemond (1977) observed amino acid N (g/day) = 32.3 DOM (kg/day) − 8.63, with r^2 = 0.90. Their organic matter intakes ranged from 4.7 to 14.6 kg/day, N intake ranged from 140 to 430 g/day, and digestible IP ranged from 11.2 to 23.1 percent of DOM in 49 observations. Rohr et al. (1979) observed amino acid N (g/day) = 31.42 DOM (kg/day) − 40.56, with r^2 = 0.85. Their organic matter intakes ranged from 8.88 to 15.14 kg/day, N intake ranged from 205 to 413 g/day, and crude protein ranged from 12.9 to 15.6 percent of dietary DM in 21 observations. These two equations indicate that DOM is a primary determinant of protein entering the small intestine. Journet and Vérité (1979) observed non-ammonia N (g/day) = 23.85 DOM (kg/day) + 0.60 *in vitro* nondegradable N (g/day) + 8.6, with R^2 = 0.886

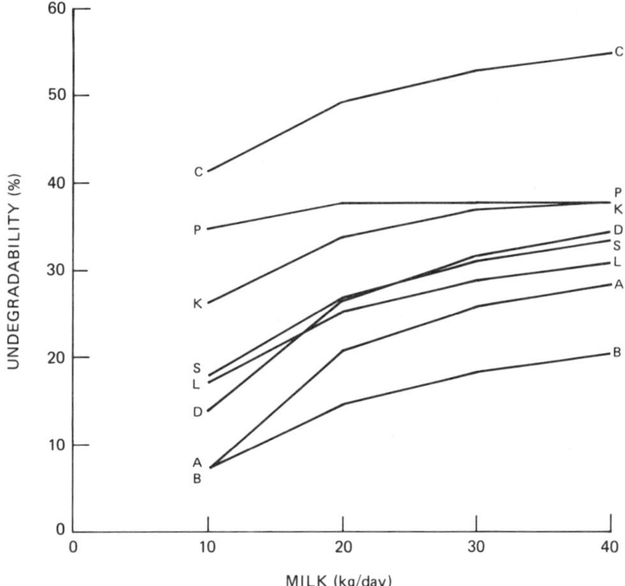

FIGURE 4   Undegradability of dietary intake protein as a function of milk production. A, ARC; B, Burroughs; C, Chalupa; D, Danfaer; K, Kaufmann; L, Landis; P, PDI; and S, Satter.

in equation 2 with lactating cows. Their DOM intakes ranged from 4.3 to 12.2 kg/day, and nondegradable N intakes ranged from 40 to 266 g/day in 42 observations. The equation of Tamminga and van Hellemond always gives a greater expectation than the equation of Rohr et al. (1979) because Tamminga and van Hellemond sampled posterior but Rohr et al. (1979) sampled anterior to the pancreatic and bile ducts.

SYSTEM COMPARISONS

First, one type of predicted flow into the small intestine of true protein (STP) was calculated as the sum of BTP plus UIP without endogenous protein for each system. Another type of predicted flow into the small intestine of crude protein (SCP) was calculated as the sum of BCP plus UIP without endogenous protein for each system. Second, three expected protein flows into the small intestine were calculated for each system based on DOM intake and UIP intake, if required, in the three equations just discussed. These two types of estimates of protein flow will be called predicted for the two former and expected for the three latter. Comparisons of the predicted protein flow, as STP, and expected protein flow, as amino nitrogen, are in Figures 6 and 7; comparison of predicted protein flow, as SCP, and expected protein flow, as nonammonia N, are in Figure 8. Predicted flows into the small intestine from the ARC, Bur-

roughs, and Satter systems were less than expected flows in all three comparisons. This difference probably results from their low AP requirement and their low IP concentration in the DM. The predicted flow in the Landis system was always less than the expected flow, and an explanation for this is not clear. The predicted flow from the NRC was highest and generally greater than expected in the two comparisons with expected flow based on DOM. This high predicted flow for the NRC system probably results from no subtraction of NCP. The predicted flow in the Danfaer model was next highest. This probably resulted from having the highest AP requirement and the highest IP concentration in the DM. The predicted flows of the Kaufmann and PDI systems were similar to the expected flows. The predicted flow for the Chalupa system was similar to the expected flow based on DOM but decreased relative to expected flow when undegradable protein became a partial basis of expectation. This difference probably resulted from the high undegradability and the low microbial protein production per unit of DOM.

Fecal Crude Protein Equivalent Excretion Relative to Crude Protein Percentage

Fecal crude protein (FP) equivalent was calculated as the sum of indigestible bacterial protein (IBP), indigestible nucleic acid crude protein (INP) equivalent, indi-

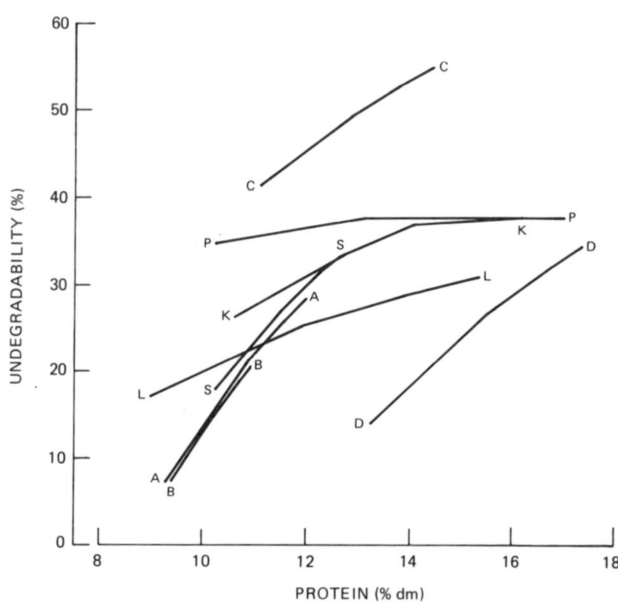

FIGURE 5   Undegradability of dietary intake protein as a function of intake protein percentage in dry matter. A, ARC; B, Burroughs; C, Chalupa; D, Danfaer; K, Kaufmann; L, Landis; P, PDI; and S, Satter.

gestible undegraded dietary protein (IUP), and fecal metabolic protein as absorbed (FPA) or net (FPN) units. The FP excretion, as a percentage of dietary IP, is expressed as a function of IP as a percentage of dietary DM (Figure 9). Reference curves are plotted from the analysis (Waldo and Glenn, 1982) of the data of Conrad et al. (1960) and Boekholt (1976). The ARC, Burroughs, Chalupa, and Satter systems and the Danfaer model predict fecal excretions lower than expected from the data of Conrad et al. (1960) or Boekholt (1976). The probable causes of these low excretions are the use of zero fecal metabolic protein in the ARC and Satter systems as a function of dietary DM and a relatively low fecal metabolic protein in the Chalupa, Danfaer, and Burroughs systems. The use of zero fecal metabolic protein produces a relatively constant percentage output of N intake in the feces, and use of a low fecal metabolic protein produces a curve with less slope than expected. The PDI system predicts a fecal output in the general range of that expected from the data of Conrad et al. (1960) and Boekholt (1976), but it declines more rapidly as concentration increases; the more rapid decline occurs because fecal metabolic protein actually decreases due to lower indigestible OM as milk production and protein concentration increase. The NRC system predicts a fecal excretion essentially equal to that expected from Boekholt

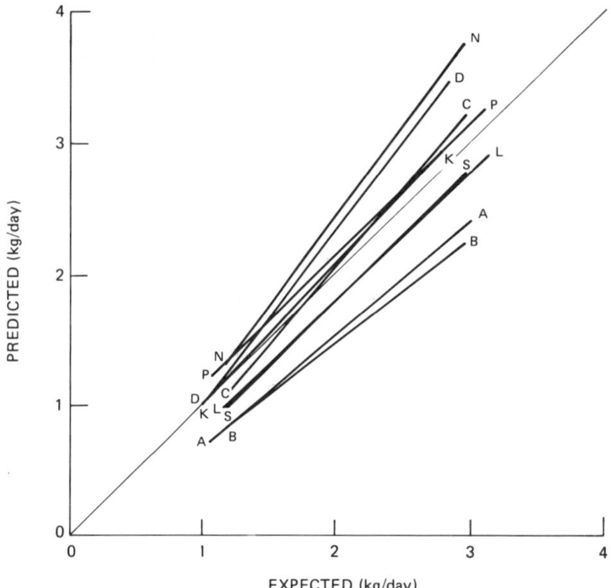

FIGURE 7   Protein flow into small intestine predicted from the system versus that expected based on the digestible organic matter of the system and the equation of Rohr et al. (1979). A, ARC; B, Burroughs; C, Chalupa; D, Danfaer; K, Kaufmann; L, Landis; N, NRC; P, PDI; and S, Satter.

(1976) and slightly higher than expected from the data of Conrad et al. (1960). The Kaufmann and Landis systems predict fecal excretions most similar to those which occur because their assumptions for fecal metabolic protein and digestibility are similar to those in the data of Conrad et al. (1960) and Boekholt (1976).

Fecal Crude Protein Equivalent Excretion Relative to Milk Production

The FP excretion, as a percentage of dietary IP, is expressed as a function of milk production in Figure 10. Two reference points for these data are 37.5 percent from Boekholt (1976) and 33 percent from the analysis (Waldo and Glenn, 1982) of the data of Conrad et al. (1960). Basically, the same comments apply to Figure 10 as were made for Figure 9. The Satter system and the ARC system, to a lesser degree, predict low outputs that are nearly constant because they assume zero fecal metabolic protein per unit of feed DM. The Chalupa, Danfaer, and Burroughs systems predict low outputs that decrease gradually with increasing milk production because they assume a minimal fecal metabolic protein. The PDI system predicts an output in the expected range, but its predicted output declines rapidly because this fecal metabolic protein output actually declines with increasing milk production. The NRC system pre-

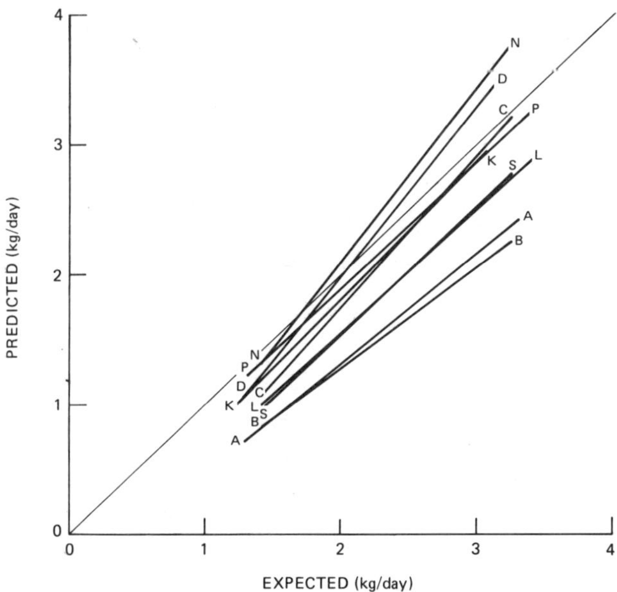

FIGURE 6   Protein flow into small intestine predicted from the system versus that expected based on the digestible organic matter of the system and the equation of Tamminga and van Hellemond (1977). A, ARC; B, Burroughs; C, Chalupa; D, Danfaer; K, Kaufmann; L, Landis; N, NRC; P, PDI; and S, Satter.

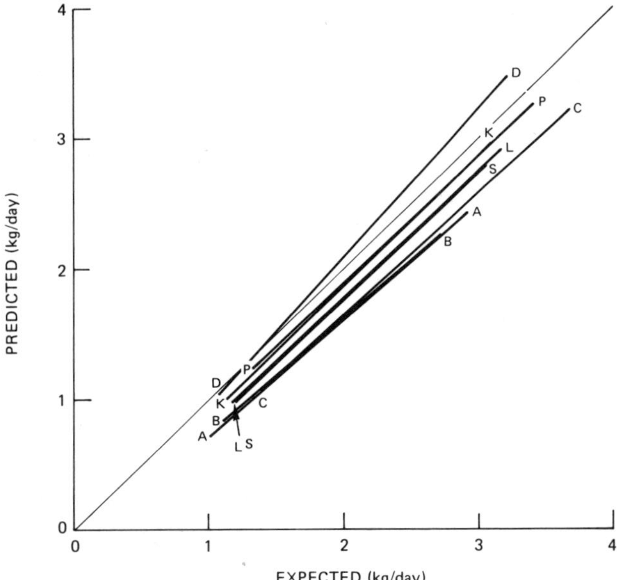

FIGURE 8   Protein flow into small intestine predicted from the system versus that expected based on the digestible organic matter plus undegraded protein intake of the system and the equation of Journet and Vérité (1979). A, ARC; B, Burroughs; C, Chalupa; D, Danfaer; K, Kaufmann; L, Landis; P, PDI; and S, Satter.

dicts more fecal output than expected because it assumes lower digestibility. The Kaufmann and Landis systems predict fecal output that follow the expected curve (Figure 9) but are slightly higher than expected.

Urinary Crude Protein Equivalent Excretion Relative to Milk Production

Urinary crude protein (UP) equivalent was calculated as the algebraic sum of rumen efflux of crude protein (REP) equivalent or a rumen influx of crude protein (RIP) equivalent; digestible nucleic acid crude protein (DNP); maintenance protein as absorbed (MPA) units that is free of any fecal metabolic N, if possible; and the protein difference of LPA minus LPN. If necessary, fecal metabolic N was subtracted to balance the system. Possibly, the tissue utilization of nucleic acids should be considered based on the finding of a 47 percent retention of activity in the tissues of the ruminating lamb by Razzaque et al. (1981).

The UP excretion as a percentage of dietary IP is expressed as a function of milk production (Figure 11). Two reference points for these data are 35.7 percent from Boekholt (1976) and 38.6 percent from the data of Conrad et al. (1960). The Satter system predicts a urinary excretion greater than expected primarily because it assumes a low efficiency of milk production. The Dan-

faer system predicts a urinary excretion greater than expected because it assumes a low efficiency for milk production and assumes an efflux of N as ammonia from the rumen to the blood. The ARC and Burroughs systems did not predict high urinary excretions as might be expected in order to balance low fecal excretions. Their predicted urinary excretions were similar to those of Boekholt and Conrad; their low dietary IP were accounted for by their high efficiencies of producing milk from AP. The PDI system is the only one that predicted an increasing UP excretion as milk production increased, and these excretions were generally lower than those of Boekholt and Conrad. This increasing urinary N excretion seems to result from the decreasing fecal metabolic N excretion at high milk production. The Chalupa system predicts UP excretions that are consistent with those of Boekholt and Conrad. The NRC system predicts low UP excretions that result from an assumption of zero DNP. The Landis system predicts a relatively low UP excretion because no DNP fraction is included. The Kaufmann system predicts a low UP excretion because the digestibility of nucleic protein equivalent is only one-half of the more common assumption.

Milk Nitrogen Output Relative to Milk Production

Output of milk protein in net (LPN) units as a percentage of dietary IP was expressed as a function of milk

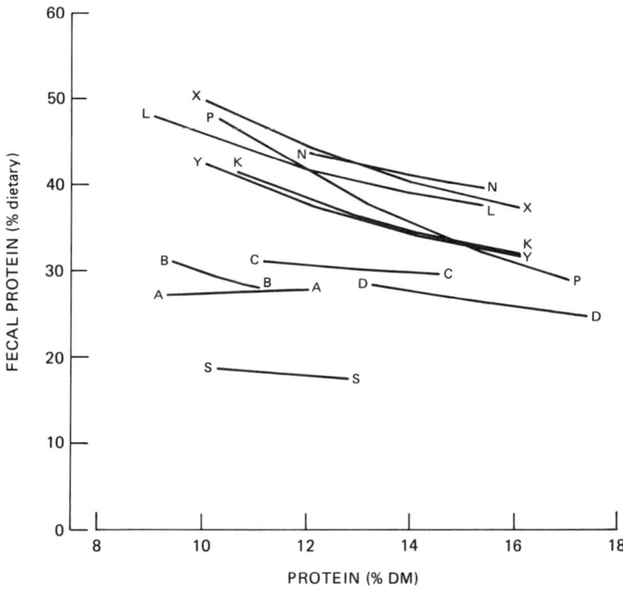

FIGURE 9   Fecal protein as a function of intake protein percentage in dry matter. A, ARC; B, Burroughs; C, Chalupa; D, Danfaer; K, Kaufmann; L, Landis; N, NRC; P, PDI; S, Satter; X, Boekholt (1976); and Y, Conrad et al. (1960).

production (Figure 12). Two reference points for these data are 24.9 percent from Boekholt (1976) and 21.7 percent from the data of Conrad et al. (1960). In all systems milk protein is assumed equal to requirement as LPN units. The high fractional output for the ARC and Burroughs systems is primarily a function of the low protein intake. All of these systems predict a higher output of dietary IP in milk than either the mean of 24.9 percent from Boekholt (1976) when mean milk production was 18.9 kg/day or the mean of 21.7 percent from the data of Conrad et al. (1960) when the mean milk production was 11.8 kg/day. Increasing milk production to the average (25 kg/day) assumed here and optimizing degradability both will increase the fractional output of dietary IP into milk. It seems overly optimistic to assume that outputs greater than 40 percent can be obtained easily.

CRITICAL COMMENTS ON OMISSIONS OF SOME SYSTEMS

Comparison and analysis of these systems as described earlier emphasize three important points that frequently are overlooked but should receive more emphasis. First, a fecal metabolic protein fraction is needed for FP excretion to correspond to *in vivo* data. Second, this fecal metabolic protein fraction should be considered either a separate component of total requirement or a feed reduction component and not be included in mainte-

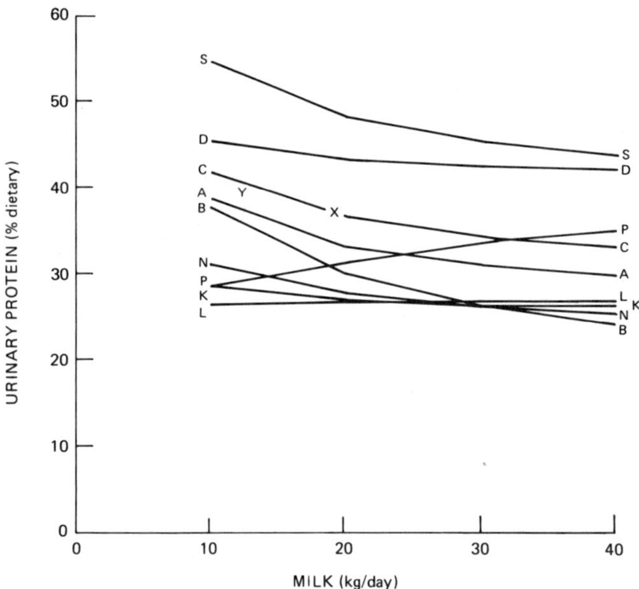

FIGURE 11   Urinary protein as a function of milk production. A, ARC; B, Burroughs; C, Chalupa; D, Danfaer; K, Kaufmann; L, Landis; N, NRC; P, PDI; S, Satter; X, Boekholt (1976); and Y, Conrad et al. (1960).

nance per se for simplicity as has been done in some cases (or ignored in others). Third, specification of the DM intake and DOM, or other energy components, are an integral part of any complete protein system.

Fecal metabolic protein is an important component of the protein requirement of the ruminant. Fecal metabolic protein represents about 57 percent of the total FP and about 20 percent of the IP requirement at 15 percent dietary IP for the negative intercept from either the equation of Boekholt (1976) or the equation (Waldo and Glenn, 1982) based on the data of Conrad et al. (1960) as its estimate. The failure to include a fecal metabolic protein factor in a protein feeding system will result in underestimation of requirements for IP percentage in dietary DM and for undegradability of dietary protein. The fraction of dietary nitrogen excreted in the feces will be underestimated, and the fractional FP excretion as a function of IP concentration will not have the characteristic *in vivo* hyperbolic curvature.

Fecal metabolic protein is most commonly related to DMI except for the PDI systems where it is related to IOMI and the NRC system where it is related to IDMI. The PDI equation (Vérité et al., 1979) is based on sheep and has an $R^2 = 0.74$; the equations of Boekholt (1976) and the data of Conrad et al. (1960) are based on lactating cattle and have $r^2 = 0.95$ and 0.92, respectively. Fecal metabolic protein is more highly correlated with DM than IOMI. Three g of FMP/100 g of DMI is a good simple interim proportion. Fecal metabolic protein is a

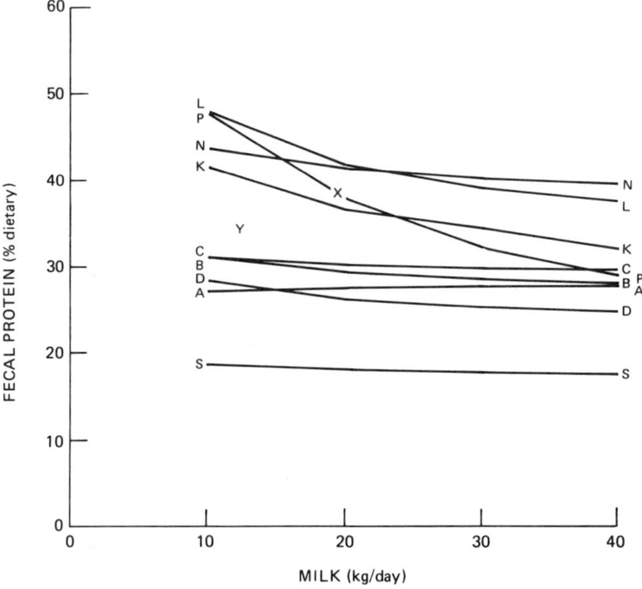

FIGURE 10   Fecal protein as a function of milk production. A, ARC; B, Burroughs; C, Chalupa; D, Danfaer; K, Kaufmann; L, Landis; N, NRC; P, PDI; S, Satter; X, Boekholt (1976); and Y, Conrad et al. (1960).

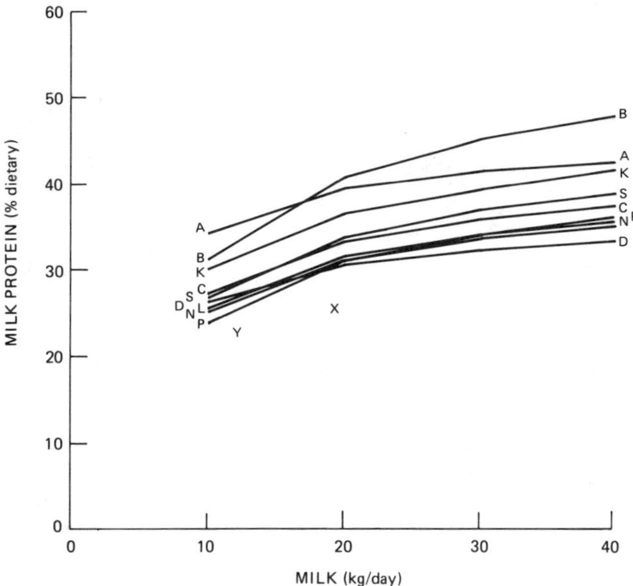

FIGURE 12   Milk protein as a function of milk production. A, ARC; B, Burroughs; C, Chalupa; D, Danfaer; K, Kaufmann; L, Landis; N, NRC; P, PDI; S, Satter; X, Boekholt (1976); and Y, Conrad et al. (1960).

function of DM intake, maintenance protein is a function of body weight, and production protein is a function of milk output. Fecal metabolic protein is considered most simply either as a separate component of the total requirement along with maintenance and production as used by Danfaer (1979), Kaufmann (1979), and Landis (1979) or as a feed reduction component as used by Burroughs et al. (1975b). When the units are considered as FPA, there is little conceptual difference between these methods of accounting. Operationally, this is much simpler than either having the fecal metabolic protein for a part of dietary DM included in maintenance and the remainder accounted for in the production requirement as in the NRC (1978) system or having one-third of the total fecal metabolic protein requirement per unit of DM in maintenance (Chalupa, 1980a).

No AP system is complete until all of the integral energy components required in the system are described. The BCP requirement per se and its contribution to the animal's need for AP are a function of the energy fermented in the rumen; generally, this component is apparently FOM. The fecal metabolic protein requirement is related to another feed component; generally, this component is dietary DM. The relationship between these two components or the digestible energy concentration in the diet thus is needed. The assumption made here is that digestible organic matter, TDN, or energy digestibility must be asymptotic at about 67 percent based on the analyses of Tyrrell and Moe (1975) of the data of Wagner and Loosli (1967). This assumption of a constant energy concentration is different from the PDI assumption where energy concentration is 85 percent greater for high milk production than for low milk production. The relative changes of concentration versus undegradability of protein are affected largely by the relative changes of digestibility and intake of energy, respectively, in requirements for higher production.

Feed Evaluation

The metabolizable or absorbable (AP) protein concept is an attempt to improve protein feeding of ruminants. It requires a descriptive separation of dietary intake protein (IP) into (1) a ruminally degraded fraction (DIP) and (2) an undegraded fraction (UIP). Ideally, this separation requires experiments with duodenally cannulated animals, techniques to separate bacterial and protozoal crude protein (BCP) and undegraded intake protein (UIP), and measurements of total protein flow. Although this remains the reference technique, its complexity has stimulated research to define simpler techniques for routine feed analysis. Tables of degradation data for various protein sources from cannulated animals are listed as Appendix Table 4.7 by ARC (1980), by Chalupa (1975a), and in Table 6 of this publication.

RUMINAL DEGRADATION ESTIMATION

Laboratory Procedures

The process of selecting simple techniques for routine feed analysis is continuing. Presently, tabulated data or several promising, but yet not generally accepted, predictive techniques are employed for various feeds. Expected degradability (DIPIP) estimates of variable sophistication and accuracy are presented in the Iowa and French protein evaluation systems. Burroughs et al. (1975b) present data on 90 common U.S. feeds. The tabular data in the French system were calculated from either solubility in salt solution or *in vitro* ammonia accumulation. Vérité et al. (1979) present data on 32 feeds, and Demarquilly et al. (1978) and Vérité and Demarquilly (1978) present data on 50 feeds. Other tabulations of solubility (Crooker et al., 1978; Waldo and Goering, 1979) and solubility and *in situ* rumen DIPIP (Crawford et al., 1978) are available.

The validity of using a single tabular value for a feed class is reduced as the variation within a feed class increases. Variation in the solubility of feeds within a class was implicit in the ranges suggested by Leng et al. (1977). The French system (Demarquilly et al., 1978; Vérité and Demarquilly, 1979) also uses ranges within feed classes based on solubility. Waldo and Goering (1979) observed ranges in insolubility of proteins in 15 feeds assayed with four methods. Ranges also are observed with dynamic techniques where four commercial samples of solvent-extracted cottonseed meal had 37.5 ± 6.6 percent (as a standard deviation) UIP and two samples of screw-press cottonseed meal had 62.6 ± 3.7 percent UIP (Broderick and Craig, 1980). In 16 samples of fish meal prepared in the laboratory (Mehrez et al., 1980), the *in situ* disappearance of nitrogen was 39.2 ± 8.6 percent. Such large variation within a feed class suggests that simple tabular values, even though estimated with systems having high predictive value, may differ greatly from the value of a specific feed.

The difficulty of obtaining UIPIP data with cannulated animals and the variation within feed classes cause the search for accurate predictive assays for individual feeds to continue. A summary of the correlations among UIPIP assays, insolubility assays, and *in vivo* responses is given in Table 5. This table includes production responses in addition to UIP passage at the duodenum. High correlations of predictive assays with production responses should encourage further consideration, but low correlations do not necessarily imply low predictive value because the animals may not have had the potential to use additional absorbed amino acids. The assays with consistently higher correlations are *in situ* bags (Gonzales et al., 1979; Stern et al., 1980), *in vitro* ammonia accumulation (Hagemeister et al., 1976; Siddons and Beever, 1977), autoclaved rumen fluid (Waldo, 1977b; Waldo and Tyrrell, 1980), and certain proteolytic enzymes (Poos et al., 1980a). Combinations of sev-

TABLE 5 Interrelationship Among *In Vivo* Responses, Undegradability Assays, and Insolubility Assays

First covariate / Second covariate	n	r^2	Reference
Milk production			
In situ bags	7	.79	Gonzalez et al., 1979
Growth			
Fungal protease	9	.71–.76	Poos et al., 1980a
Bromelain	9	.25–.49	Poos et al., 1980a
Ficin	9	.49–.55	Poos et al., 1980a
Papain	9	.34–.44	Poos et al., 1980a
Bacterial protease	9	.46–.58	Poos et al., 1980b
Burroughs	9	.32	Poos et al., 1980b
Sodium chloride, .15 M	9	.17	Poos et al., 1980b
Sodium hydroxide	9	.49	Poos et al., 1980b
Water, hot	9	.55	Poos et al., 1980b
Tissue nitrogen deposition			
Autoclaved rumen fluid	4	.95	Waldo and Tyrrell, 1980
Nitrogen retention, balance			
Autoclaved rumen fluid	4	.91	Waldo and Tyrrell, 1980
Autoclaved rumen fluid	6	.99	Waldo, 1977b
Burroughs	4	.28	Wohlt et al., 1976
Burroughs (forage component)	3	.04	Sniffen, 1974
Burroughs (concentrate component)	3	.85	Sniffen, 1974
Burroughs (total diet)	3	.59	Sniffen, 1974
In vivo protein degradation			
In situ bags	3	.61	Stern et al., 1980
In vitro nitrogen digestion	11	.44	Siddons and Beever, 1977
In vitro ammonia accumulation	11	.61	Siddons and Beever, 1977
In vitro ammonia accumulation	10	.63	Hagemeister et al., 1976
Pronase	11	.10	Siddons and Beever, 1977
Pepsin	11	.56	Siddons and Beever, 1977
Pepsin	21	.27	Siddons et al., 1976
Duodenal N flow/feed N			
Pepsin	5	.96	Beever et al., 1976
In situ bags			
Autoclaved rumen fluid	28	.29	Crawford et al., 1978
Burroughs	28	.44	Crawford et al., 1978
Sodium chloride	28	.22	Crawford et al., 1978
In vitro ammonia accumulation			
Autoclaved rumen fluid	6	.01–.34	Crooker et al., 1978
Burroughs, modified	6	.01–.61	Crooker et al., 1978
Burroughs	6	.01–.49	Crooker et al., 1978
Burroughs	25	.98	Henderickx and Martin, 1963
McDougall's	6	.02–.64	Crooker et al., 1978
Sodium chloride	6	.04–.45	Crooker et al., 1978
Sodium hydroxide	7	.27	Little et al., 1963
Water	7	.14	Little et al., 1963
Autoclaved rumen fluid			
Burroughs, modified	7	.01	Crooker et al., 1978
Burroughs	7	.04	Crooker et al., 1978
Burroughs	28	.61	Crawford et al., 1978
Burroughs	350	.53	Waldo and Goering, 1979

TABLE 5 Continued

First covariate / Second covariate	n	r^2	Reference
McDougall's	7	<.01	Crooker et al., 1978
Sodium chloride, .15 M	7	.65	Crooker et al., 1978
Sodium chloride, .15 M	27	.69	Crawford et al., 1978
Sodium chloride, .15 M	350	.37	Waldo and Goering, 1979
Sodium hydroxide	7	.07	Little et al., 1963
Water, hot	350	.59	Waldo and Goering, 1979
Water	7	.04	Little et al., 1963
Burroughs, modified			
Burroughs	7	.82	Crooker et al., 1978
McDougall's	7	.97	Crooker et al., 1978
Sodium chloride, .15 M	7	.02	Crooker et al., 1978
Burroughs			
McDougall's	7	.82	Crooker et al., 1978
Sodium chloride, .15 M	46	.86	Crooker et al., 1978
Sodium chloride, .15 M	27	.82	Crawford et al., 1978
Sodium chloride, .15 M	350	.66	Waldo and Goering, 1979
Water, hot	350	.41	Waldo and Goering, 1979
McDougall's			
Sodium chloride, .15 M	7	<.01	Crooker et al., 1978
Water, hot	350	.24	Waldo and Goering, 1979
Sodium hydroxide			
Water	7	.12	Little et al., 1963

eral procedures such as solubility and *in situ* (Zinn and Owens, 1983) may be helpful.

Some special problems have been observed in the analysis of certain feeds by some techniques. The French system is generally based on solubility in salt solution for dry feeds and pressed juice from wet fermented feeds (Vérité and Demarquilly, 1978). Vérité et al. (1979) observed that such solubility data were generally well correlated with *in vitro* DIPIP as measured by ammonia accumulation as earlier observed by Henderickx and Martin (1963). But solubility was lower than expected for cereals, soybean meal, and beet pulp and higher than expected for horse beans and peas based on the general relationship to degradability. Entrapped liquids may cause some protein solubility estimates to be misleading. The DIPIP for corn gluten meal by *in situ* bag technique was 14 percent, but by duodenally cannulated animals it was 45 percent (Stern et al., 1980). This difference occurred because it formed a viscous mass in the bags. *In situ* results will vary due to pore size and thoroughness of washing.

Animal Procedures

Two assay procedures are being used that use animals for more than *in vitro* or *in situ* fermentations. Klopfenstein et al. (1982) use a cattle growth assay to determine the value of supplemental proteins relative to soybean meal. A basal or negative control diet contains supplemental nitrogen as 100 percent urea while the reference or positive control diet contains supplemental nitrogen as 40 percent from soybean meal and 60 percent from urea. Test proteins are substituted for soybean meal, and a protein efficiency is calculated as the incremental gain from the protein supplement divided by the incremental protein intake from that protein supplement. Relative values are calculated by dividing the protein efficiencies of test proteins by the protein efficiency of soybean meal. This is a useful transitional method, but its general use as an assay would tend to ignore the increasing evidence for the large variation among lots within a feed class and effects of dietary energy level and food intake on ruminal degradation of protein.

Danish researchers (Möller and Thomsen, 1977) use duodenally cannulated animals and regression techniques to estimate UIP and BCP production relative to the DM ingested. A protein source is fed at different nitrogen percentages in the feed dry matter, X, and the ratio of duodenal nitrogen to feed nitrogen, Y, is related to X by the hyperbolic regression equation, $Y = a + b/X$. The constant, a, is interpreted as the fraction of feed protein escaping degradation in the rumen. The constant, b, is interpreted as microbial nitrogen fixation into protein per 100 g of dry matter ingested. Variation within feed classes and the complexity of experiments with cannulated animals make the use of this method unlikely as a general assay.

Since milk production responds rapidly to changes in protein status, direct use of lactation response to assay the need for additional UIP appears feasible and directly applicable. Calderon Cortes et al. (1977) abruptly changed the IP protein fed to ewes at the start of the third week of lactation to 77 percent and again at the start of the fourth week to 106 percent of that fed in the second week. The corresponding changes in milk production were 83 percent and 101 percent. The corresponding changes in milk protein output were 76 percent and 101 percent. Note that milk production and milk protein output responded rapidly to both the square wave decrease and increase in crude protein fed. The milk production responses to short-term changes in UIPIP observed by Gonzales et al. (1979) also support this hypothesis.

PROTEIN INDIGESTIBILITY

At least five common feeds may contain sizable portions of their protein in bound or indigestible form. These feeds are hay-crop silages (Goering et al., 1974), dehydrated alfalfa (Goering, 1976), citrus pulp (Am-

merman, 1973), and corn distillers dried grains and brewers dried grains (Waldo and Goering, 1979). The Cornell system considers that acid detergent insoluble nitrogen (Goering and Van Soest, 1972) is bound and indigestible (Van Soest et al., 1982). Pepsin insoluble nitrogen is another possible method for determining this fraction. Heat and chemicals that decrease the ruminal degradation of proteins can increase the amount of bound protein. The bound and indigestible fraction must be subtracted from the undegradable fraction since it does not contribute absorbable amino acids.

PROTEIN FRACTIONS AND DEGRADATION

Protein degradation has been described as a function of time when using *in vitro* and *in situ* fermentations or proteolytic enzymes. Most of these data fit a general model with three pools or fractions:

A—NPN or protein that is degraded very rapidly;
B—protein that is degraded at a rate similar to the rate of passage (0.02 to 0.07 h^{-1}; and
C—bound or unavailable protein that is degraded very slowly.

Theoretically, each pool or fraction has a degradation rate that is assumed to be fractional, that is, a constant proportion of the residue is degraded per unit of time. The fractional degradation rates are:

k_{dA}—fractional degradation rate for A that may be in the order of 10 times greater than the rate of passage;
k_{dB}—fractional degradation rate for B that may be between 10 times and one-tenth the rate of passage; and
k_{dC}—fractional degradation rate for C that may be in the order of one-tenth the rate of passage.

Practically, k_{dA} is usually considered infinite and A is considered to be entirely degraded; k_{dC} is usually considered zero and C is considered to be entirely passed. Only B is usually considered to be affected by the relative rates of passage, k_{pB}, and k_{dB} at any time (see p. 215 of Bray and White, 1966). The fraction of B that is degraded will be $k_{dB}/(k_{dB} + k_{pB})$ and the fraction of B that is passed will be $k_{pB}/(k_{dB} + k_{pB})$. The fraction of total protein that is degraded,

$$D = A + k_{dB}B/(k_{dB} + k_{pB}),$$

and the fraction of total protein that is passed,

$$P = k_{pB}B/(k_{dB} + k_{pB}) + C.$$

As fraction C is often an asymptotic residue, it may or may not relate to bound and unavailable fraction discussed above. A time lag for bacterial attachment and penetration by ruminal fluid may precede degradation.

Pichard and Van Soest (1977) used proteolytic enzymes to describe subfractions B_1 and B_2 plus their fractional rate constants. Soluble fraction A and unavailable fraction C were estimated independently by chemical assay. The implicit fractional rate constant for A was infinity and no time lags were implied. Van Soest et al. (1982) extended the system to include subfraction B_3, for some proteins. Broderick and Craig (1980) used *in vitro* rumen fermentations to describe a biexponential system of A and B plus their fractional rate constants. Unavailable fraction C was implicitly zero and no time lag was implied. Broderick (1982) suggested that the biexponential might be simplified by using a Michaelis-Menten approach where degradation rate = Vmax/ Km. The estimated proportions escaping the rumen based on Michaelis-Menten degradation rates were similar to proportions based on biexponential degradation rates for casein and unheated cottonseed meal.

Schoeman et al. (1972) used *in situ* bags to measure protein degradation at 12 or 24 h, and later Mehrez and Ørskov (1977) used synthetic fiber (normally dacron or nylon) bags *in situ* for determining the degradation of protein in the rumen at several times. Ørskov et al. (1980) presented a detailed description of this *in situ* technique and its application. Mohamed and Smith (1977) used the *in situ* technique to describe a fraction A that was washed out of the bag and a fraction B plus its fractional degradation rate. Fraction A was 1 minus the antilog of the intercept value, and its fractional rate was assumed to be infinity. Neither fraction C nor a time lag were considered. Nocek et al. (1979) calculated fractional degradation rates from 0 to 2 h and from 2 to 12 h for concentrates or 2 to 48 h for forages. Their first rate applies to fraction A, and the second rate applies to fraction B of the general model. Pool sizes for fraction A and fraction B are not explicitly defined. No fraction C was considered and a time lag of 2 h for the 2 to 12 h or 2 to 48 h degradation rates was implied by default. Grummer and Clark (1982) calculated fractional degradation rates from 0 to 1 h, 1 to 4 h, and 4 to 16 h. The first rate applies to fraction A, and the third rate applies to fraction B of the general model. Pool sizes for fraction A and fraction B were not explicitly defined. No fraction C was considered, and a time lag of 4 h for the 4 to 16 h degradation rate is implied. Zinn et al. (1981) described fraction A as that lost at 4 h *in situ* and calculated fractional degradation rates from 4 to 12 h and from 12 to 24 h. The first rate thus applies to fraction B_1, and the second rate applies to fraction B_2 of the general model. Pool sizes for fraction B_1 and B_2 were not explicitly calculated, and no fraction C was considered. A time lag of 4 h is implied for fraction B_1 and a time lag of 12 h is im-

plied for fraction B_2. A termination time of 12 h is implied for fraction B_1. Owens and Zinn (1982) described fraction A independently by solubility due to washout of small particles through the pores of the dacron bags and calculated fractional degradation rates for the residue, so explicit and implicit assumptions are the same as described for Zinn et al. (1981).

The definition of rates without the simultaneous explicit definition of pools is not the proper way to apply differential equations to biological systems. Certain treatments of feeds may change the pool size, while others change the fractional rate. Examples of the former are reduction of lignin by either chemical or genetic methods that affect the pool of potentially digestible fiber more than it affects its fractional rate of digestion (Waldo and Jorgensen, 1981). The pool sizes of the protein fractions varied among feedstuffs (Krishnamoorthy et al., 1982). Knowledge of pool sizes, degradation rates, and passage rates are needed to quantitate protein degradation in the rumen. Secondly, time lags are an occasional component of descriptive biology using differential equations and are consistent with the assumption of fractional rate constants. But termination times are inconsistent with the concept of fractional rate constants. Conceptually, fractional degradation continues for infinite time. Choice of a termination time similar to mean retention time of particles in the rumen may leave a protein residue in dacron bags similar to amounts of protein escaping *in vivo* ruminal degradation (Zinn and Owens, 1983) but do not provide values for modeling to other passage rates.

Ørskov and McDonald (1979) combined data from *in situ* degradation rate measurements with independent data on rate of passage using chromium labeled protein. They calculated an effective percentage degradation, $D = A + [k_{dB}B/(k_{dB} + k_{pB})][1 - e^{-(k_{dB} + k_{pB})t}]$, where t is time after feeding. This effective percentage degradation is the amount of protein degraded at any time, t, when both passage and degradation are possible such as in the rumen. They calculated A as the intercept and considered the possibilities of a fraction C and a lag time but did not use them. McDonald (1981) included a lag time and relaxed the constraint that A + B = 1. Stern et al. (1983a) combined rates of degradation and passage using either the procedure of Ørskov and McDonald (1979) for the final value or the procedure of Miller (1980), where degradation, $D = A + k_{dB}B/(k_{dB} + k_{pB})$. Neither fraction C nor a lag time is considered. Erdman (1982) combined rates of degradation and passage to calculate the protein degradation, $D = A + k_{dB}B/(k_{dB} + k_{pB})$ or protein passing, $P = k_{pB}B/(k_{dB} + k_{pB}) + C$. The implicit fractional rate for A is instantaneous, and fraction C is included but no lag time is considered.

Krishnamoorthy et al. (1983) described an *in vitro*

technique for estimating rumen proteolysis using protease from *Streptomyces griseus*. Krishnamoorthy et al. (1983) compared estimates of ruminal escape protein using *in vitro* proteolysis and the *in situ* bag technique for 12 concentrate mixtures when assuming the rate of passage to be $0.04\ h^{-1}$. The *in vitro* proteolysis estimates of escape protein were more highly correlated ($r^2 = 0.61$) with *in vivo* escape protein than *in situ* estimates of escape protein were correlated ($r^2 = 0.41$) with *in vivo* escape protein.

All of the models described above imply two simultaneous first-order processes operating on each pool and are subject to the same criticisms about the validity of these assumptions as used by Baldwin et al. (1977a) about a similar model of fiber degradation and passage. The concept of pools and rates is avoided by the use of summative incremental models (Kristensen et al., 1982; Stern et al., 1983a; Stern and Satter, 1983). Such models do not provide detailed analytic insight since they do not consider pool sizes, rates, or lag times.

An interim proposal for a system of describing the degradation of feed proteins seems to require three fractions. Fraction A is assumed to be instantaneously degraded. An intermediate degradable fraction, B, is assumed to degrade at a fractional degradation rate that makes the extent of degradation a function of residence time. The relative proportions of B degraded and passed depend on the relative rates of degradation and passage. Fraction C is assumed to have a zero rate of degradation and must pass undegraded. A time lag may be considered for B by conventional techniques of differential equations where t minus the time lag rather than t, per se, is used. Conceptually, these three fractions and the rate of degradation of B could be estimated from time series data obtained from *in vitro* and *in situ* fermentations or proteolytic enzymes. Proper selection of the initial time should allow a mathematical definition of fraction A as 1 minus the intercept value using theory of differential equations and nonlinear estimation much as done by Mohamed and Smith (1977) or Ørskov et al. (1980). Proper selection of the termination time should allow a mathematical definition of fraction C as an asymptotic value using theory of differential equation and nonlinear estimation. This leaves fraction B = 1 − (A + C). Simplification by use of a single B fraction and single rate constant should be thoroughly considered. The Michaelis-Menten approach of Broderick (1982) and the stochastic modeling approach of Matis and Tolley (1980) are possibilities. Such a system seems a reasonable compromise because J. H. Matis as cited by Broderick (1982) has suggested that about 10 time points are required to define each fraction and its fractional degradation rate.

The theoretical enzymatic and chemical arguments for a larger number of subfractions are valid but the cost and difficulty of quantitation rapidly increase. Matis and Tolley (1980) point out the statistical difficulty of estimating more than two fractions and suggest that a deterministic model evaluated at the average rate will always underestimate the mean of the corresponding stochastic model. When one considers that a dairy ration will generally contain at least three feed components—forage, grain, and protein supplement—it may be well to heed the comments of Matis and Tolley (1980), that ". . . a small, simple stochastic model can often be substituted for a large, complex deterministic model. . . ."

Protein degradation estimates from *in situ* bags are very sensitive to the conditions of measurement (Mohamed and Smith, 1977). Degradation of cottonseed meal was reduced when *in situ* bags were incubated in a host animal that received a concentrate diet compared to a forage diet (Owens and Zinn, 1982). Mohamed and Smith (1977) found steers fed an 85 percent corn diet had no difference in the soluble fraction but a reduced rate of degradation to one-third as compared with sheep fed an alfalfa hay diet. Ørskov et al. (1983) found degradation greater in sheep fed grass than in sheep fed barley and no consistent difference between cattle and sheep fed grass. Mohamed and Smith (1977) also found a threefold increase in rate of degradation when the host animal was adapted to the protein being tested. Zinn et al. (1981) and Owens and Zinn (1982) used a reference protein of soybean meal in an attempt to control variation and observed a high correlation between predicted and observed protein bypass. Evidence is accumulating that rates of degradation decline as feed intake increases and that this reduction is greater for proteins that have higher rates of degradation at low feeding levels (Erdman, 1982). This finding implies that the fractional rates are not constant and that protein degradability differences will be reduced at high intakes or low pH. Decreasing particle size increased both the soluble fraction and the fractional rate of degradation (Mohamed and Smith, 1977), but Ehle et al. (1982) have not observed any increase in degradation rate with decreasing particle size. Fine soybean meal increased ruminal bypass of nitrogen compared to coarse soybean meal (Netemeyer et al., 1980) but did not alter rumen ammonia, blood urea, total tract digestibility, and milk production. Heating of soybean meal decreased both the soluble fraction and the fractional rate of degradation to about one-third that of an unheated control (Mohamed and Smith, 1977). Mixed total diets showed large deviation from that expected based on the ingredients (Nocek et al., 1979).

Degradation of Dietary Crude Protein in the Reticulo-Rumen

INTRODUCTION

Intake protein (IP) that passes to the omasum is often called "bypass" or "undegraded" protein (UIP) to differentiate it from protein synthesized by microbes (BCP) in the rumen and from endogenous secretions. These terms can be confusing and overlapping. The IP that passes to the omasum consists of two fractions. These are: (1) protein that resists microbial attack in the rumen; and (2) protein that evades attack in the rumen and passes to the omasum without thoroughly mixing with ruminal contents. Protein flushed out of the rumen at feeding time and passing through the esophageal groove would fall into this category. The term "undegraded" protein is most suited to the first fraction, while "bypass" would be more suited to the second fraction. Measurements *in vitro* usually attempt to quantitate "undegraded protein," while *in vivo* measurements include both fractions. The BCP synthesized in the rumen, UIP, and endogenous protein together total the amount of protein entering the omasum.

Rumen microorganisms cause major transformations of dietary nitrogenous compounds. Most forms of nonprotein nitrogen are converted almost quantitatively to ammonia. True protein is degraded to a variable extent to peptides and amino acids in the rumen. Peptides and amino acids are utilized for synthesis of BCP, or are further hydrolyzed and deaminated, producing ammonia as the major end product, which contains N.

Although rumen microbes may supply 60 to 80 percent of the amino acids (protein) absorbed from the intestine (AP), much interest has been focused on the amount of UIP. Medium- to high-producing ruminants rely on some IP escaping degradation in the rumen since the quantity of BCP is inadequate to support high rates of growth, wool production, or milk production. The proportion of UIP must increase as production levels increase, using feeds and technology of the present. The supply of UIP can be a limiting factor at high levels of animal performance. This was illustrated in the work of Hogan and Weston (1967), which stimulated much research in N utilization by ruminants in the following decade. Insights gained during the last decade form the basis for much of the following discussion.

MECHANISM OF PROTEIN DEGRADATION

This topic has been reviewed by Tamminga (1979). Therefore, discussion will be limited to an overview of some of the major features. The IP entering the reticulo-rumen may be degraded by both bacteria and protozoa. Degradation involves basically two steps: (1) hydrolysis of the peptide bond (proteolysis) to produce peptides and amino acids; and (2) deamination and degradation of amino acids. Russell et al. (1983) suggest that the hydrolysis of peptides to amino acids is the rate-limiting step. Free amino acid concentrations in ruminal ingesta are normally extremely low (Annison et al., 1959; Lewis, 1962), suggesting that proteolysis is normally the rate-limiting step in protein degradation. This view is supported by Nugent and Mangan (1978, 1981).

The proteolytic enzymes appear to be associated primarily with the bacterial cell wall with a small amount of cell-free activity probably resulting from cell lysis (Allison, 1970). An example is the protease produced by the rumen anaerobe *Bacteroides amylophilus*. This protease is present on the outer cell surface and hydrolyzes protein extracellularly (Blackburn, 1968; Blackburn and Hullah, 1974). Proteolytic enzymes are associated with many rumen bacteria, and proteolytic activity of rumen microorganisms is not greatly altered by diet

(Blackburn and Hobson, 1962; Allison, 1970). As discusssed later, diet can have an effect on protein degradation in the rumen, perhaps indirectly through altering pH and bacterial numbers or types.

Protease activity appears to be "trypsin-like" in nature. Craig and Broderick (1984) observed that when casein was incubated *in vitro* with rumen microorganisms, losses of lysine and arginine were disproportionately large. Stern and Satter (1982) reported similar results with *in vivo* studies. Craig (1981) observed that the artificial trypsin substrate benzoylarginine ethyl ester inhibited *in vitro* casein degradation, but synthetic substrates for chymotrypsin had little effect. These results imply that bacterial proteases may be trypsin-like in activity, preferentially exposing lysine and arginine residues to further degradation by microbial exopeptidases and deaminases. This suggests that use of a trypsin inhibitor may reduce ruminal protein breakdown and improve utilization of feed protein.

Following proteolysis, liberated peptides or amino acids may leave the reticulo-rumen, be utilized for microbial growth, or be degraded to ammonia and fatty acids. Amino acids are rapidly degraded in the rumen, and therefore only small quantities of free amino acids would be available for absorption or passage from the reticulo-rumen. The half-life of eight essential amino acids incubated with strained rumen fluid was 2 h or less (Chalupa, 1976).

MEASURING PROTEIN DEGRADATION

Measuring protein degradation by rumen microbes is a difficult task. There can be wide variation in protein degradation within and among feedstuffs, as well as significant differences among animals with regard to rumen environment and retention time of feed in the reticulo-rumen. There are many sources of analytical error, the most important of which is distinguishing between BCP and UIP. Considerable caution must be exercised in applying the results of a single experiment, and replication of experiments or studies is necessary to help identify contributing variables. No single technique or experimental design is fully adequate at the present time.

Despite the difficulties of making *in vivo* measurements of protein degradation, *in vivo* measurements are essential, for they serve as the standard against which all chemical or *in vitro* methods for estimating protein degradation must be evaluated. Chemical or *in vitro* methods for estimating protein degradation are important for screening or monitoring purposes, but they must be validated *in vivo* and must not serve as the only estimate of protein degradation.

In Vivo *Methods*

In vivo measurements are usually performed with surgically prepared animals equipped with cannulae in the rumen and abomasum or small intestine.

Determination of digesta flow with a reentrant cannula may be accomplished with total collection of the ingesta, or more commonly by use of an indigestible digesta marker and collection of spot samples (Zinn et al., 1980). When using animals prepared with T-type cannulae, spot samples are taken and flow rate of digesta is calculated by reference to digesta markers. Although *in vivo* measurements of protein flow to the intestine must be the primary source of information about protein degradation in the rumen, it must be recognized that measurement of digesta flow to the duodenum is subject to considerable error. Digesta markers currently used are not ideal markers and do not always reflect the solid or liquid phase that they are intended to represent. The use of digesta markers to measure flow to the small intestine has been reviewed (Faichney, 1975, 1980; Warner, 1981).

The amount of UIP can be estimated as the difference between IP and the sum of endogenous and BCP entering the abomasum or small intestine. Procedures for estimating BCP are available, utilizing microbial markers such as nucleic acids, diaminopimelic acid (DAPA), aminoethylenephosphonic acid (AEP), or one of the radioisotopes, ^{35}S, ^{32}P, or ^{15}N (Clark, 1977). Estimates of BCP based upon digesta or microbial marker techniques are subject to errors inherent in those techniques. In practice, some investigators use microbial markers present in bacteria only and therefore do not include protozoal protein in their estimates. Protozoal protein can be important under certain feeding conditions (Harrison and McAllan, 1980). Estimates of endogenous protein are variable and difficult to obtain. Consequently, endogenous protein is often ignored, leading to an overestimate of UIP when difference techniques are used. The extent of this error probably is not large.

Another approach to estimate the amount of UIP is available. This method is based on the increase in flow of protein to the small intestine in response to incremental additions of IP (Stern and Satter, 1982). Unfortunately, this technique is useful only with feeds having a relatively high protein content. It assumes that protein content in the ration does not influence the measurement in question (Zinn et al., 1980).

In Situ *Method for Estimating Protein Degradability*

While the use of cannulated animals can provide estimates of protein degradation in the rumen, *in vivo* estimates are labor intensive and time consuming. Alternative techniques that can provide rapid, yet reasonable estimates of protein degradation for a wide variety of feedstuffs are desirable. Unfortunately, alternative techniques tested to date have one or more major limitations. One of the more promising approaches is the dacron or nylon bag technique. Mehrez and Ørskov (1977) suggested that this *in situ* technique is suitable for determining degradation of protein. The simplest application of the *in situ* technique for estimating protein degradation is to suspend the bag in the rumen for an arbitrary period of time, thus giving a relative estimate of protein degradation. Alternatively, the extent of protein degradation can be determined at the moment when a predetermined percentage of the truly digestible organic matter has disappeared from the dacron bag, thus simulating the extent of digestion in the rumen of normally fed animals (Ørskov and Mehrez, 1977). Unfortunately, ruminal retention time and ruminal organic matter digestion vary among diets, intake levels, and many other conditions.

Several methods have been used to combine *in situ* N disappearance and ruminal dilution rate information (Ørskov and McDonald, 1979; Mathers and Miller, 1981; McDonald, 1981; Stern and Satter, 1982). The first three methods are similar in approach and use rate constants for both nitrogen disappearance and passage rate. The procedure applied by Mathers and Miller involves the following:

$$\text{Fraction of protein degraded} = A + k_{dB}B/(k_{dB} + k_{pB}),$$

where the terms in the equation are as previously described.

It may be inappropriate to apply a single rate constant to the degradation of that portion of protein remaining in the bag after the soluble protein has disappeared. Several rate constants are probably involved with most feedstuffs, depending upon the number and amount of each type of protein present. Rate constants for digestion of N usually have more influence on protein degradation than rate constants for passage from the rumen (Ørskov and McDonald, 1979; Mathers and Miller, 1981). The following example illustrates this point. Many protein supplements, and most feedstuffs, will have a ruminal passage rate (k_{pB}) within the range of 0.03 to 0.07 h^{-1} (Ganev et al., 1979; Hartnell and Satter, 1979; Lindberg, 1982; Stern and Satter, 1982). Using an arbitrary value of .1 for k_{dB} and 0.3 for A, protein degradation would decrease from 0.84 to 0.71 as the rate constant (k_{pB}) for passage of undigested residue from the rumen increased from 0.03 to 0.07 h^{-1}. This is a rather modest change in degradation as a result of a large change in rumen retention time. The value used for k_{dB} will determine, of course, how much influence

k_{pB} will have on protein degradation. Manipulations that increase k_{pB}, such as increased feed intake, will have their greatest effect when k_{dB} is small (Mathers and Miller, 1981).

Stern and Satter (1982) have described a more empirical approach for combining *in situ* N disappearance and ruminal dilution rate. Protein degradation is obtained by summing the product of protein remaining in the rumen (determined in a rate of passage study) and the fractional disappearance of N from the dacron bag at 10 different time intervals. The approach is analogous to the method of Castle (1956) for calculating mean retention time of digesta in the gastrointestinal tract. The approach avoids reliance on a single rate constant for describing the rate of protein degradation.

The *in vitro* bag technique is subject to variables that can influence the estimate of protein degradation. Pore size of the cloth can influence the rate and extent of N disappearance from the bag. Entry of feed particles and colonization of bag contents by rumen bacteria can lead to an underestimate of protein degradation, and variation in the washing technique can lead to error. Although the *in situ* bag technique is an imperfect and empirical approach, it incorporates animal and microbial factors helpful in quantitating protein degradation in the rumen.

Protein Solubility as a Means of Estimating Protein Degradation

Soluble proteins tend to be more rapidly or completely degraded (Hendrickx and Martin, 1963). Unfortunately, some segments of the feed industry have assumed that soluble protein is degraded in the rumen and insoluble protein is not. Early reports of animal work, often quoted to relate protein solubility with protein degradation in the rumen, reveal no basis for equating soluble protein with degradable protein and insoluble protein with undegradable protein, except for extreme examples such as zein and casein (McDonald, 1952; Chalmers et al., 1954; el-Shazly, 1958; Tagari et al., 1962; Little et al., 1963; Whitelaw and Preston, 1963; Tagari, 1969).

Soluble proteins are generally more vulnerable to proteolysis than insoluble proteins. Accessibility of proteins to proteases is much greater if the protein is in solution. It seems likely, however, that some feed protein can be hydrolyzed directly from the solid state without an intervening soluble stage, similar to the digestion of cellulose. Relatively insoluble proteins such as zein can be extensively degraded in the rumen, given adequate time. It may be that much of the protein hydrolysis is occurring on the surface of the feed particle.

Large differences exist between soluble proteins in the

rate at which they are hydrolyzed. Nugent and Mangan (1978) studied the degradation of casein, fraction I leaf protein, and bovine serum albumin *in vitro* using sheep rumen fluid. All three proteins were buffer soluble but differed greatly in the rate at which they were hydrolyzed (casein > fraction I leaf protein > bovine serum albumin). Treatment of bovine serum albumin with dithiothreitol, which breaks some of the disulfide bonds cross-linking the protein, caused a substantial increase in its rate of rumen proteolysis. It was concluded that differences in the rates of microbial hydrolysis of these proteins were caused by structural and not solubility differences. Mahadevan et al. (1980) further examined this question by incubating soluble and insoluble proteins with partially purified protease from *Bacteroides amylophilus*, one of the principal proteolytic organisms in the rumen. Their results showed that serum albumin and ribonuclease A, both of which are buffer soluble, were relatively resistant to hydrolysis, and that buffer-soluble proteins from soybean meal, rapeseed meal, and casein were hydrolyzed at different rates. Interestingly, buffer-soluble and -insoluble proteins of soybean meal were hydrolyzed at almost identical rates. Mahadevan et al. (1980) concluded that buffer solubility of a protein is not an indication of susceptibility to hydrolysis by rumen bacterial protease.

A somewhat different perspective relating protein solubility and susceptibility to proteolysis has been discussed by Pichard and Van Soest (1977). They concluded from protein solubility and proteolysis studies that there are four general categories of N in ruminant feeds. Fraction A is a water-soluble NPN fraction containing primarily nitrate, ammonia, amines, and free amino acids. Insoluble fractions include a rapidly degradable protein fraction B1, a more slowly degradable protein fraction B2, and an unavailable fraction C.

Application of this approach to partitioning of N in silages might have potential. Fermentation of forages increases the amount of N in fraction A and may increase the amount in fraction C if the forage has undergone heat damage. Whether a combination of solubility and protease information can be used to predict *in vivo* protein degradation of forages remains to be demonstrated. Since some proteins are soluble in water, it would appear that fraction A for some feedstuffs would contain true protein in addition to nonprotein nitrogen.

Stern and Satter (1982) evaluated the relationship between N solubility in Burroughs mineral buffer (Burroughs et al., 1950), N disappearance from dacron bags, and *in vivo* measurements of degraded intake protein (DIP) for 34 total mixed diets containing various dietary N sources. They found that N solubility was highly correlated with N disappearance from bags in the rumen for short exposure times, but as exposure time increased

the correlation between these procedures progressively decreased, to be expected due to the dynamics of degradation. Correlation between *in vivo* crude protein degradation and N disappearance from dacron bags became significant at 12 h of rumen exposure and increased to 0.68 at 24 h. The correlation between N solubility and degradation of protein *in vivo* was only 0.26, indicating that solubility may be a poor predictor of protein degradation, when the dynamics of ruminal passage are not taken into account. Solubility of a protein varies with the solvent used (Crooker et al., 1978), and care is required in interpretation of some experimental results. Measurements of protein solubility were described by Wohlt et al. (1973) and Waldo and Goering (1979).

Information on solubility of proteins in buffers is presently being used in France for estimating rumen protein degradation in a protein evaluation scheme known as the PDI system (Vérité et al., 1979). Protein solubility in mineral buffer and, in some cases, ammonia production *in vitro* (Vérité and Demarquilly, 1978; Vérité and Sauvant, 1981) are related to published information on flow of N from the rumen in sheep and cattle fitted with omasal, abomasal, or intestinal cannulae. They concluded that, on average, all of the soluble dietary crude protein and 35 percent of the insoluble dietary crude protein were degraded in the rumen. The hazard in using such a constant is discussed above.

Several feed companies in the United States are presently using information on protein solubility in mineral buffers to formulate dairy rations. One group of companies formulates diets to provide not less than 15 percent and not more than 25 percent of the total dietary protein as soluble protein, and has claimed that such formulations increase milk yields (Braund et al., 1978).

In Vitro *Ammonia Production for Estimating Protein Degradation*

A common approach for estimating protein degradation involves incubation of the test protein with rumen fluid and measurement of subsequent ammonia production. The chief advantage of this method is its simplicity, provided a source of ruminal ingesta is available. The method has several disadvantages, however, that limit its usefulness. Microbial growth and ammonia uptake occur simultaneously with protein degradation and ammonia release. This frustrates quantitative measurements by making it difficult to equalize microbial growth across a variety of feedstuffs. Broderick (1978) has attempted to overcome this problem by inhibiting deamination and uptake of amino acids by the microbes, thus enabling a direct measure of proteolysis. Another problem is that incubation conditions (substrate, end products, pH) in a batch culture change with time, so rates of both ammonia production and uptake change.

In vitro ammonia production has been the principal method used for obtaining estimates of protein degradation for the ARC system of protein evaluation (Roy et al., 1977) and has been used in France with the PDI system mentioned earlier. Vérité et al. (1979) comment that *in vitro* incubation with rumen digesta is probably superior to solubility for estimating protein degradation, but that the procedure is not suitable for routine analysis. They feel that there is satisfactory agreement between the two methods for most feedstuffs, but that protein solubility gives lower estimates of degradation for cereals, soybean meals, and sugar beet pulps and higher estimates for horse beans and peas. With these feedstuffs, *in vitro* ammonia production estimates, rather than solubility estimates, were used, and these were termed "corrected solubility" values in their feedstuff tables. It appears that when *in vitro* ammonia production and protein solubility give similar estimates of protein degradation for a class of feeds, solubility information is used. When agreement is not good, *in vitro* ammonia production is used.

In summary, estimation of ruminal protein degradation in the rumen is a complex problem. A primary difficulty, *in vivo*, *in situ*, and *in vitro*, is to distinguish between microbial and dietary protein. Secondly, all of the laboratory or *in vitro* procedures for estimating *in vivo* protein degradation have one or more major flaws that can invalidate the estimates. It therefore appears necessary to continue with the tedious and costly *in vivo* experiments with cannulated animals to determine protein degradation of the major feedstuffs. These determinations are also subject to error, and considerable replication is advised. Equally important with the *in vivo* studies are the feeding variables (cited earlier) that can influence protein degradation in the rumen.

Protein solubility, or *in vitro* methods, will continue to be the source of protein degradation estimates for the minor feedstuffs for the near future, even though these "short-cut" procedures are potentially misleading. On the other hand, protein solubility or *in vitro* methods can be used to monitor changes within a feedstuff or to screen similar feedstuffs. For example, Beever et al. (1976) found a high negative correlation ($r^2 = 0.96$) between soluble N in perennial ryegrass, determined as the N soluble after incubation with 0.01 percent pepsin in 0.1 N HCl for 16 h, and the quantity of total nitrogen entering the small intestine. The solubility and, in this case, the degradation of the ryegrass protein were altered by drying at different temperatures and by formaldehyde treatment. It is reasonable to expect solubility or the *in vitro* methods to predict differences in protein degradation more accurately when applied to a group of similar feedstuffs that when used across a diverse group of feedstuffs that differ in physical and chemical properties.

EXTENT OF PROTEIN DEGRADATION IN THE RUMEN

Both ruminant nutritionists and livestock producers seek more quantitative information on the extent of protein degradation in the rumen. Three of the metabolizable protein systems that have been proposed to replace crude or digestible protein for ruminants are dependent upon protein degradation values (Burroughs et al., 1975a; Roy et al., 1977; Vérité et al., 1979). Table 6 and Appendix Table 2 contain a list of feedstuffs and estimates of the percentage of crude protein that escapes destruction in the reticulo-rumen. All of these estimates were obtained with sheep and/or cattle at various feed intakes and having abomasal or duodenal cannulae. It is clear from these tables that (1) estimates of the amount of protein escaping degradation in the reticulo-rumen are extremely variable, and (2) *in vivo* information is deficient or nonexistent for many feedstuffs of major dietary importance. Part of the variation in degradation estimates is due to analytical error associated with the *in vivo* measurements and part to variation in feedstuffs, the diets used and amounts fed, the experimental animals employed, and method of feeding and physical nature of the diet. The values for protein degradation in Table 6 must be used with caution. In some instances the values reported in Table 6 do not agree with other *in vivo* information on protein degradation where the feedstuff in question was part of a mixed diet, but where degradation of individual protein sources was not reported.

Most evidence suggests that the small grains, such as barley and oats, have protein that is more degradable than the protein in corn. Soybean meal protein is a relatively degradable protein. *In vivo* information on whole cottonseeds and cottonseed meal is very limited, but cottonseed meal prepared by the expeller process may be more resistant than that prepared by the solvent process (Broderick and Craig, 1980; Goetsch and Owens, 1985).

Many by-product feeds appear relatively resistant to ruminal degradation. Brewers grains, distillers grains, corn gluten meal, fish meal, blood meal, and meat and bone meal are more resistant than most feed grains and oil meals. Up to 50 percent or more of the protein in these feedstuffs escapes degradation.

The protein in most forages is quite susceptible to degradation. The *in vivo* estimates of protein degradation in forages are variable, reflecting not only the difficulty

TABLE 6   Tentative Estimates of Undegraded Protein for Common Feedstuffs When Total Dry Matter Intake Is in Excess of 2 Percent of Body Weight

Feed	Number of Measurements	Mean Fraction of Undegraded Protein	Standard Deviation
Feedgrains			
Barley	2	0.21	0.07
Corn	3	0.65	0.06
Sorghum grain	8	0.52	0.15
Oil meals			
Cottonseed meal (solvent)	6	0.41	0.12
Cottonseed meal (prepress)	2	0.36	0.02
Cottonseed meal (screw press)	2	0.50	0.07
Linseed meal	1	0.44	—
Peanut meal	2	0.30	0.08
Rapeseed meal	1	0.23	—
Soybean meal	10	0.28	0.14
Sunflower meal	2	0.24	0.05
By-product feeds			
Blood meal	1	0.82	—
Brewers dried grains	5	0.53	0.14
Corn gluten meal	3	0.55	0.06
Distillers dried grains	2	0.62	0.07
Fish meal	4	0.80	0.12
Meat meal	1	0.76	—
Meat and bone meal	2	0.60	0.11
Forages			
Alfalfa hay	4	0.28	0.08
Alfalfa (dehydrated)	3	0.62	0.04
Bromegrass hay	2	0.32	0.12
Corn silage	1	0.27	—
Timothy hay	2	0.42	0.11

in measuring degradation of low-protein feedstuffs but also the wide variation in forage protein content due to harvesting and method of preservation. Since forages provide the bulk of protein in many ruminant rations, more quantitative information on degradation of forage protein is needed.

It has been assumed in studies with protein degradation that the individual amino acids are equally susceptible to degradation. There may be preferential hydrolysis of some amino acids from the peptide or protein molecule. Secondly, free amino acids may differ in their rates of degradation. Chalupa (1976) addressed the latter question and noted that amino acids differ markedly in their rates of degradation by rumen microbes. Arginine and threonine were rapidly degraded. Lysine, phenylalanine, leucine, and isoleucine formed an intermediate group, while valine and methionine were least rapidly degraded. Nevertheless, all free amino acids

were rapidly catabolized. Stern et al. (1983b) measured the relative loss of individual amino acids from protein in intestinally cannulated lactating dairy cows receiving incremental amounts of corn gluten meal. The six most degraded amino acids in descending order were lysine, isoleucine, histidine, arginine, valine, and leucine. The basic amino acids appear more extensively degraded than acidic amino acids (Stern and Satter, 1982). This is different from the studies with free amino acids (Chalupa, 1976).

Besides knowing the extent of ruminal protein degradation for feedstuffs, it is important to know the relative value of protein sources for supporting animal production. With this in mind, Klopfenstein (1980) used the slope-ratio technique (Hegsted and Chang, 1965) for evaluating protein sources for growing ruminants. This approach should reflect not only on the amount of protein that escapes ruminal degradation, but also the quality and availability of protein that escapes the reticulo-rumen. Klopfenstein (1980) calculated a "protein efficiency value" for nine different feedstuffs, with soybean meal being assigned an efficiency value of 100 percent. All other feeds were evaluated relative to soybean meal, depending on growth in beef cattle. The protein efficiency values for blood meal (ring dried), blood meal (conventional), corn gluten meal, brewers grains, dehydrated alfalfa, meat meal, distillers grains and distillers grains (with solubles) were 250, 200, 190, 190, 185, 173, and 137 percent, respectively. Although *in vivo* measurements of protein degradation were not made, these animal growth data support the concept that the protein supplements tested were more resistant to ruminal degradation than soybean meal. Lactating ewes have been used to evaluate protein sources of differing extents of degradation, and the correlation coefficient between milk yield and degradation of the protein supplement was -0.89 (Gonzalez et al., 1979). Similar information with lactating dairy cows is needed.

The ultimate test of any nutrient or feedstuff is how well it supports animal production. Obviously protein degradation information is important, but care must be exercised in relying too heavily on data that are not quantitative and tell only part of the story. Animal response data are essential in the final evaluation of protein supplements.

FACTORS INFLUENCING PROTEIN DEGRADATION IN THE FORESTOMACH

The extent to which protein is degraded in the rumen will depend upon microbial proteolytic activity in the rumen, microbial access to the protein, and rumen turnover. Differences in the proteolytic potential of rumen digesta under a variety of feeding conditions have been small. Microbial access to the protein seems to be the most important factor influencing protein degradation in the rumen.

Tertiary Structure of the Protein

The three-dimensional structure of protein is important in determining whether the protein will be degraded or not. For example, ovalbumin is slowly degraded because it is a cyclic protein having no terminal amino or carboxyl groups (Mangan, 1972). Proteins with extensive cross-linking, such as disulfide bonds, are less accessible to proteolytic enzymes and are relatively resistant to degradation (Nugent and Mangan, 1978). Examples of such proteins are hair and feathers. Proteins treated with formaldehyde have methylene cross-linking and are normally degraded to a lesser extent (Ferguson et al., 1967). These and other features of protein structure dictate the vulnerability of protein to hydrolysis in the rumen.

Proteins in feeds are composed primarily of four types: albumins, globulins, prolamines, and glutelins. Albumins and globulins are usually more soluble than prolamines and glutelins (Sniffen, 1974). This is unfortunate because albumins and globulins usually have higher biological values than prolamines and glutelins.

Rumen Factors

Retention time of feed protein in the rumen can influence protein degradation. Proteins retained for a short time are degraded to a lesser extent than those with a longer retention time. Ruminal retention time of dietary ingredients varies among animals (Balch and Campling, 1965), among species (Church, 1969), and among diet ingredients (Hartnell and Satter, 1979). Retention time is influenced by particle size of the feed (Balch and Campling, 1965) and by the quantity of feed eaten (Balch and Campling, 1965; Zinn et al., 1981; Lindberg, 1982). A comprehensive review of factors influencing digesta passage through the gut is available (Warner, 1981). The amount of UIP in lactating cows eating either 8.2 or 12.9 kg of dry matter daily was 29 and 45 percent, respectively (Tamminga et al., 1979). High-producing ruminants consuming large quantities of feed are likely to have a larger percentage of UIP than animals consuming low or moderate amounts of feed. The effect of level of intake on retention time of feed particles, however, is sometimes quite small (Hartnell and Satter, 1979; Varga and Prigge, 1982), and the impact on protein degradation may often be minor (Miller, 1973) or without effect (McAllan and Smith, 1983).

A summary of studies showing the relationship between level of feed intake and retention time in the rumen reveals that intake has a large effect on ruminal retention time when daily intake is less than approximately 2 percent of body weight. The decrease in ruminal retention time associated with increased feed intake is much diminished when feed consumption is in excess of approximately 2 percent of body weight (Prange, 1981). A somewhat similar observation by Alwash and Thomas (1971) indicated that mean retention time of forage particles was related to the log of feed intake. In conclusion, increased feed intake can reduce protein destruction in the rumen, but the influence of feed intake on residence time in the rumen and therefore on protein degradation is diminished as intake is increased. Level of feed intake may have some effect on protein degradation aside from influencing residence time. For example, lower rumen pH, which usually accompanies higher levels of feed intake, may reduce bacterial activity and proteolytic activity. As pointed out earlier, variation between protein sources in rate of protein degradation is greater than variation in retention time in the rumen and would therefore have more impact on extent of total proteolysis in the rumen.

Increasing the dilution rate of rumen fluid has been demonstrated to increase the flow of protein from the rumen of sheep (Harrison et al., 1975) and steers (Cole et al., 1976b; Prigge et al., 1978). Part of this increase is probably due to a net increase in BCP (Harrison et al., 1975; Harrison and McAllan, 1980) and part due to an increase in the amount of UIP (Hemsley, 1975). Rumen fluid dilution rates have been increased by feeding or by ruminal infusion of artificial saliva, sodium bicarbonate, or sodium chloride.

Environmental temperatures can influence the residence time of feed in the rumen. Kennedy et al. (1976) demonstrated that sheep in a cold environment had an increased rate of digesta passage. This increased BCP and the amount of UIP. In a more recent study, Kennedy et al. (1982) found that the percentage of UIP in the rumen increased from 20 to 24 percent for alfalfa hay and from 40 to 49 percent for bromegrass hay when sheep were exposed to cold temperatures. No effect of temperature on extent of protein degradation of a barley-canola seed meal diet was observed. The turnover time (h) of [103]Ru-phenanthroline in the rumen for the alfalfa, bromegrass, and barley-canola meal diets at warm and cold temperatures were: 18.4, 12.3; 15.6, 15.3; and 38.9, 32.3.

Feeding of monensin has been shown to reduce dietary protein degradation *in vitro* (Whetstone et al., 1981) and in the rumen (Poos et al., 1979b; Isichei and Bergen, 1980). Although the amount of information is limited, it appears that UIP can be increased by approximately one-third with monensin feeding. However, monensin may inhibit BCP synthesis (Chalupa, 1980b), resulting in little or no net increase in total protein supply to the intestine.

Rumen pH could affect protein degradation by altering microbial activity and by changing protein solubility. Rumen pH is normally between 5.5 and 7.0, and proteins with an isoelectric point in this range would have altered solubility and possibly altered protein degradability. Also, fiber may limit microbial access to plant protein, and reduced fiber digestion at a lower pH might be involved as well (Ganev et al., 1979).

Proteolysis and deamination are affected by pH, but experimental results are conflicting. As reviewed by Tamminga (1979), the bulk of evidence suggests that the optimum pH for both proteolysis and deamination is between 6 and 7. There are reports of lower pH optima for ruminal proteases and deaminases, but since activity of both will be largely dependent upon total bacterial numbers, rumen pH in a range between 6 and 7 should be compatible with maximum microbial activity. Under most feeding situations, pH in the rumen is in a range where extensive breakdown of dietary protein can occur.

Little is known about the effect of ammonia concentration on proteolysis or deamination. Since the main pathway of ammonia fixation by rumen bacteria may differ according to the prevailing concentration of ammonia (Erfle et al., 1977), it might be suggested that catabolic processes in rumen bacteria are influenced by ammonia concentration. For example, ammonia, through end product inhibition, might alter the rate of protein hydrolysis. Nikolic and Filipovic (1981), however, were not able to demonstrate an effect of ammonia concentration on the degradation rate of corn protein. Very low ammonia concentrations would affect total proteolytic activity to the extent that ammonia might limit total microbial growth (Poos et al., 1979a).

Feed Processing and Storage

Many feedstuffs are exposed to heat during processing or storage. By-product feeds are frequently produced by an aqueous extraction process and are often dried for marketing. This exposure to heat can make the protein more resistant to degradation (Ferguson, 1975). Ensiled feeds may experience elevated temperatures for a sustained period of time, resulting in more resistant protein (Merchen and Satter, 1983b).

Feed processing methods such as pelleting, extrusion, and steam rolling and flaking may generate enough heat to alter feed protein. In terms of optimum protection of protein, however, it is likely that more heat is required than most commercial processing methods will provide. Moisture level, quantity of soluble carbohydrate present in the feedstuff, maximum temperature, and time-temperature relationships are some of many factors that will influence the effects of feedstuff exposure to heat (Goering and Waldo, 1978). Heat treatment of feeds to reduce protein degradation in the rumen has potential (Beever and Thomson, 1981), and quantitative information is needed.

Protection produced by heating can be counterbalanced by decreases in total tract digestibility and biological value. The Maillard reaction between sugar aldehyde groups and the free amino groups of protein is responsible for much of the heat damage to protein when reducing sugars are present. However, proteins can be damaged by reactions other than Maillard type. Condensation reactions make essential amino acids nutritionally unavailable (Ferguson, 1975).

Beever et al. (1981) noted that pelleting a mixture of Italian ryegrass and timothy, which had been dried at high temperature, reduced degradation of dietary protein from 69 to 47 percent. The effect of pelleting in this experiment may be due to heat or to an influence on retention time of the forage in the rumen. Pelleting demonstrates how changing the physical form of a feedstuff can influence protein degradation.

Ensiling of feeds can convert large portions of true protein into NPN (Bergen et al., 1974; Goering and Waldo, 1974). This may lower the amount of protein potentially available for passage from the rumen. Formation of NPN is particularly evident with silages of high moisture content (Merchen and Satter, 1983b). However, other factors (including hydration rate) may influence these events.

Chemical treatment of feedstuffs has been used to provide partial protection against breakdown in the rumen. Feeding trials with formaldehyde-treated casein appeared very promising (Ferguson et al., 1967), and extensive experiments with formaldehyde treatment of forage have been conducted. Presently, formaldehyde-treated feeds are used in Europe. Although treatment of commercial protein supplements with formaldehyde has been disappointing, a combination of formic acid-formaldehyde has been used to assist preservation of direct-cut forages (Waldo, 1977a). This process is also employed in Europe.

Tannins have been used to protect protein from degradation in the rumen. Driedger and Hatfield (1972) reported that addition of 10 percent tannin to soybean meal fed to lambs increased rate and efficiency of gain and nitrogen retention and decreased *in vitro* deamination by 90 percent. The high level of tannin used in those experiments would appear not to be practical. Isopropanol, propanol, and ethanol have been used to increase the resistance of protein in soybean meal to degradation by rumen microbes (Van der Aar et al., 1982). Inhibitors of amino acid deamination in the rumen have been tested (Chalupa and Scott, 1976).

There is great potential for protecting feed protein from excessive destruction and loss in the rumen. One of the major advantages of feeding protected protein would be greater opportunity for utilization of NPN for BCP synthesis in the rumen and the economy inherent with NPN use. A balance is needed, however, in the amount of UIP and the amount of dietary nitrogen made available for BCP synthesis. Much remains to be learned about practical ways to alter protein degradation in the rumen. For more complete summaries of experimental work, the reader is referred to Chalupa (1975a), Clark (1975a), Ferguson (1975), Huber and Kung (1981), and Owens and Bergen (1983).

Microbial Growth

INTRODUCTION

Microbial flow from the rumen can meet 50 percent or more of the amino acid requirements of ruminants in various states of production (Ørskov, 1982). Therefore, it is important to understand the total rumen microbial ecology and factors affecting it.

Microbial growth is a pivotal point in any ruminant protein system. There is an optimum balance between requirements for microbial growth and substrate availability. The optimum is dictated, in part, by utilization of degraded protein and carbohydrate from any of the feedstuffs or ingredients used in diets. Protein degradation in the rumen, in some cases, exceeds carbohydrate availability, and protein wastage occurs. In others, the reverse is true, and digestion of carbohydrate in the rumen is reduced.

Generalized schemes of carbohydrate and protein degradation have been presented (Russell and Hespell, 1981), and Figure 1 contains an overall protein scheme. The focus for defining microbial growth is to understand the substrate being fermented, the organisms fermenting this substrate, and microbial requirements.

MAJOR CLASSES OF ORGANISMS IN THE RUMEN

Hungate (1966) and Russell and Hespell (1981) have described the rumen microbial genera. The rumen ecological system is complex and not entirely understood. The population is diverse with interdependence of various types of organisms (Meers, 1973).

Rumen bacteria can be divided into three major classes based on substrate affinity: cell wall digesters, general (those that can digest both cell wall and cell contents) digesters, and cell contents digesters. The last two categories may include those bacteria that can be classed as secondary fermenters, i.e., those that utilize substrate from primary fermenters.

Russell and Hespell (1981) recently outlined the distribution of fermentation niches and major fermentation products of some major rumen bacteria. Few bacterial species have proteolytic capability, and a few species are responsible for most of the digestion of cellulose. As a result, the composition of the diet can alter the rumen ecology and influence microbial growth, total microbial mass, and extent of dry matter digestion. In general, an increase in any one component of the substrate, particularly nonstructural carbohydrate, could result in proliferation of the digesting organism, usually at the expense of other species.

Protozoa are divided physiologically into two major subclasses: flagellates and ciliates (Hungate, 1966; Russell and Hespell, 1981). Flagellates occur in young calves shortly after feeding but decrease as animals age. The major protozoal population in adults is ciliate, which is subdivided into two major groups: holotrichs and oligotrichs. Holotrichs are relatively simple, similar to paramecia, (Hungate, 1966; Russell and Hespell, 1981) and usually belong to the *Dasytricha* or *Isotricha* genera. Oligotrichs are more complex with surface projections, cilia, and skeletal plates. Example species are *Entodinia* and *Diplodia* (Hungate, 1966). Their role in the rumen is poorly understood. They engulf bacteria and feed particles and may influence proteolysis and recycling of bacterial nitrogen (Leng and Nolan, 1984) and delay starch metabolism (Hungate, 1966).

Bacterial populations are usually in the range of 10^7–10^9/ml of rumen fluid and protozoa are 10^2–10^6/ml. Since individual protozoa mass may be 10^3 times that of bacteria, the total ruminal mass of protozoa in the rumen may equal that of bacteria.

BACTERIAL NUTRIENT REQUIREMENT

Bacterial growth can be rapid (doubling times range from 14 minutes to 14 hours), and the rate is a partial function of the availability of substrate at any given time interval (Bergen et al., 1982). Their nutrient requirements are complex and dynamic and are a function of the microbial maintenance requirement as well as the requirement for growth (Russell and Hespell, 1981).

Growth for animals is normally described as change in mass per unit of time. At steady-state conditions in the rumen, bacteria grow or multiply at a rate only sufficient to replace those passing out of the rumen or lysing, since at steady state the population of cells remains constant. Growth rate, as measured by turnover of isotope labels, is an index of the rate at which cells are replaced, and gross yield from the rumen is the multiple of the replacement rate and the population in the rumen. Net yield, in contrast, is the multiple of dilution rate of microbes and population in the rumen. The difference is cell lysis. Yield also is commonly calculated as the multiple of substrate(s) use and $Y_s \cdot Y_s$ is yield per unit of substrate fermented. This can be further fractioned into a Y_s^{max} times substrate minus a maintenance coefficient times the population. If yield equals Y_s and also equals population times dilution rate, then population equals Y_s dilution rate. If dilution rate and Y_s are constant, as under steady-state conditions, then population becomes a function of substrate available per unit of time. The amount of ATP per unit of substrate fermented may differ with different types of bacteria. ATP yield and the maintenance coefficient and turnover rate determine the efficiency of microbial growth.

Microbial cell composition has been demonstrated to vary considerably (Hungate, 1966; Hespell and Bryant, 1979), depending on many factors, including microbial type, growth phase, and rate of nutrient availability. Table 7 illustrates some of the variation in cell composition of bacteria in different metabolic states. The high-polysaccharide-content group represents those species that grow at slow rates and may be inefficient due to energetic uncoupling (Hespell and Bryant, 1979). Uncoupling may occur in the rumen of animals fed at or below maintenance, fed low-nitrogen diets high in noncell wall material, or with diurnal fluctuations of limiting nutrients, especially protein (Hespell and Bryant, 1979; Cotta and Russell, 1982). The Table 7 high-protein composition represents bacteria in the rapid growth phase, specifically bacteria that have adequate substrate and nutrient supply. These data are *in vitro* and caution should be used in their application.

Nutrient requirements could be expressed in terms of rate of microbial growth and microbial type. It is difficult to consider bacterial type because of the range in maintenance requirement and variation in nutrient requirements.

Carbohydrate

Microbes can be classified according to substrate specificity (Russell and Hespell, 1981). The microbial mass can be divided into two major categories: primary and secondary fermenters (Van Soest, 1982). The primary fermenters degrade the complex cell wall, starch, and sugars. The secondary fermenters utilize the products produced by the primary group. Cell yield may not parallel the amount of carbohydrate fermented (Hespell and Bryant, 1979; Russell et al., 1983) when factors necessary for growth are absent or when some factor increases the maintenance cost.

The major available carbohydrate fractions of plant cell wall are cellulose, hemicellulose, and pectin. Noncell wall carbohydrates are primarily starch, fructosans, and sucrose. Insoluble and partially unavailable cellulose and hemicellulose constitute from 15 to 66 percent of most diets of ruminant animals. Although it is a part of the cell wall, pectin along with the soluble carbohydrate is rapidly and completely fermented, while starch is the primary insoluble storage carbohydrate that is susceptible to rumen escape.

The objective in feeding ruminants is to obtain a rate of digestion of the complex carbohydrate substrate to maximize nutrient intake and availability of nutrients from the rumen and the lower tract. Digestion is maximized in an ecosystem balanced in acidity, nutrient availability, and fermentation products both within and among microcolony niches. Due to methane and heat losses that accompany fermentation in the rumen, energetic efficiency may favor small intestinal over ruminal digestion of nutrients, but certain nutrients are poorly or not digested in the small intestine.

TABLE 7   Composition of Microbial Cells[a]

	General	High Polysaccharide Maintenance/ Low Turnover	High Protein Lipid 4 × Maintenance/ High Turnover
Protein	47.5	47.5	65.0
RNA	11.4	8.0	8.0
DNA	3.4	1.0	1.0
Lipid	7.0	7.0	12.0
Polysaccharide	12.3	30.1	7.6
Peptidogylcan	14.0	2.0	2.0
Ash	4.4	4.4	4.4

[a]Hespell and Bryant (1979).

Protein or Nitrogen

Microbial nitrogen requirements vary qualitatively. Many fiber digesters require ammonia and may require branched chain C_4 and C_5 acids for protein synthesis and growth (Hungate, 1966; Johnson and Bergen, 1982; Russell and Sniffen, 1984). Amino acids appear mildly stimulatory to a few organisms such as *Ruminococcus albus*, *R. flavefaciens*, and *Megasphera elsdenii* (Bryant and Robinson, 1963; Hungate, 1966; Maeng and Baldwin, 1975; Maeng et al., 1975; Leibholz and Kellaway, 1979; Russell et al., 1983). The starch, sugar, and secondary fermenters also require ammonia. However, there are several species such as *Streptococcus bovis* for which amino acids and possibly short peptides are essential (Cotta and Russell, 1982). Amino acids and branched chain volatile fatty acids are required by cellulolytic bacteria *in vitro*, but crossfeeding can meet this need in the rumen under most circumstances (Hume, 1970; Stewart, 1975; Chalupa, 1976; Russell et al., 1979).

The amount of ammonia required for microbial growth has been researched, modeled, and reviewed extensively (Nolan et al., 1972; Thomas, 1973; Satter and Roffler, 1975; Smith, 1975, 1979; Mehrez et al., 1977; Baldwin and Denham, 1979; Kennedy and Milligan, 1980; Schaefer et al., 1980; Beever et al., 1980, 1981; Black et al., 1980–1981; Kang-Meznarich and Broderick, 1981). Mehrez et al. (1977) suggested that an ammonia concentration of 20 to 22 mg NH_3-N/100 ml rumen fluid was needed to maximize rate of barley dry matter fermentation. Lower values are suggested to be adequate by other workers based on *in vitro* data. (Satter and Slyter, 1974) and by Ørskov (1982) for highly fibrous diets. Poos et al. (1979a) suggested that maximum digestion and intake depend upon larger fluxes of ammonia because of greater quantities of fermentable organic matter in dairy cows fed total mixed rations. It is suggested that the requirement for ammonia is directly related to substrate availability, fermentation rate, microbial mass, and yield (Hespell and Bryant, 1979; Russell et al., 1983). Methyl amine may also play a role in ammonia uptake by microorganisms (Hill and Mangan, 1964).

Vitamins and Minerals

Vitamin requirements have been outlined by Hungate (1966) and others (Scott and Dehority, 1965). Generally, many of the organisms require biotin, PABA, thiamin, folic acid, and riboflavin. Recent results would suggest that nicotinic acid may, under certain conditions, improve the efficiency of microbial growth (Bartley et al., 1979; Schaetzel and Johnson, 1981). However, these studies need corroboration. Crossfeeding should supply the B vitamins necessary for bacterial growth in most feeding conditions.

Mineral requirements have commonly been considered for only the host animal with the exception of sulfur and cobalt. Bacterial requirements can be large (Ammerman and Miller, 1974; Spears et al., 1978), especially when one considers the requirements in terms of the dynamic microbial growth (Thomson et al., 1977). It is important that minerals like phosphorus (2 to 6 percent of cell dry matter) and sulfur for synthesis of sulfur amino acids (Hume and Bird, 1969) be available during rapid microbial growth.

PROTOZOA

Protozoa in the rumen ecosystem consume particulate cellulose, peptides, starch (which delays fermentation of non-cell wall constituents), and bacteria (Coleman, 1975; Delfosse-Debusscher et al., 1979; Demeyer and Van Nevel, 1979; Vogels et al., 1980).

Protozoa have a division time of about 15 h. If the environmental condition in the rumen is such that there is a high rumen turnover or the coarse particulate matter of the upper layer in the rumen is reduced, such as through the feeding of fine particle substrate, the population will be reduced through washout (Whitelaw et al., 1972). Although as much as 50 percent of the microbial protein in the rumen may be in protozoa, they only constitute 20 to 30 percent of the microbial nitrogen flowing to the small intestine, which may be mostly the small ciliate protozoa (Leng and Nolan, 1984). Oldham (1984) suggests that at higher levels of intake in dairy or beef cattle where the particle size in the rumen is smaller, due to a smaller particle size in the diet, and solid and liquid turnover is greater, there could be increased washout of protozoa and subsequently smaller rumen populations under typical feeding situations. If, however, animals are fed continuously, this would not be the case. Protozoa are sensitive to pH, and if rumen pH is outside the range of 5.5 to 8.0 (optimum pH 6.5) for extended periods, these populations will be reduced (Hungate, 1966).

Holotrichs rapidly assimilate soluble sugars that are stored as starch. In contrast, entodiniomorphs ingest starch and particulate matter. There is evidence that entodiniomorphs can digest cellulose, although this activity may be the result of residual enzymes produced by consumed bacteria.

Protein requirements are met variously by ingestion of peptides, preformed protein, amino acids, and, to a small degree, ammonia or possibly urea (Hungate,

1966). Protozoa and some bacteria are actively proteolytic and will digest protein and release ammonia.

The nutrient requirements of protozoa are poorly defined. It could be assumed that the requirements are proportional to composition. Research is needed in this area.

SPIROCHETES

Spirochetes have recently been characterized in the rumen (Paster and Canale-Parola, 1982). They have been found to vary from 10^5 to 4×10^6 cells/ml of rumen fluid. Thirteen strains were characterized. They were shown to utilize hydrolysis products of plant polymers. They do not ferment amino acids. It was concluded that these organisms do contribute to the breakdown of plant polysaccharide material.

FUNGI

Fungi have also been recently identified in the rumen (Bauchop, 1981; Akin et al., 1983) as having a significant role in fiber digestion. Bauchop (1981) suggests that the concentration of fiber in the ration is positively correlated with fungal concentration. It was demonstrated that the fungi preferentially colonized the lignified cells of blade sclerenchyma (Akin et al., 1983).

Further studies are needed with spirochetes and fungi to determine nutrient requirements (Akin et al. [1983], have shown a positive sulfur response by fungi), the interaction with the bacterial and protozoal mass, the dietary and environmental conditions under which they thrive, and their significance in the extent of organic matter digestion in the rumen.

MICROBIAL GROWTH AND FLOW

Microbial growth will be discussed in three contexts: microbial efficiency, microbial mass, and microbial flow. Efficiency and mass are dependent on the specific substrate available for fermentation in the rumen, pattern, composition and rate of substrate availability, and environmental factors. Microbial flow is dependent on rumen volume/passage and particle size relationships.

Most reviews of microbial efficiency have considered Y_{ATP} (microbial cells/mole ATP), protein or N/unit of fermentable organic matter, or mole of hexose fermented (Hespell and Bryant, 1979; Smith, 1979; Stern and Hoover, 1979; Steinhour and Clark, 1982). These terms are most appropriate in chemostats and possibly studies conducted with small particle diets fed frequently (Hungate, 1966).

The factors affecting microbial efficiency are numerous and complex and are beyond the scope of this discussion. Reviews and discussions of concepts and equations have been presented elsewhere (Bauchop and Elsden, 1960; Pittman and Bryant, 1964; Pirt, 1965; Forrest and Walker, 1971; Stouthamer, 1973, 1979; Stouthamer and Bettenhaussen, 1975; Hespell and Bryant, 1979; Roels, 1980; Bergen et al., 1982). Growth and its limitations can first be defined in terms of the maintenance requirement (Pirt, 1965). The maintenance requirement varies (Hespell and Bryant, 1979) among various bacteria. The impact on net microbial growth can be significant. The maintenance requirement is the net diversion of energy and/or carbon from growth-limiting (or energy-generating) substrate to processes not resulting in an increase of cell mass. The maintenance requirement is both growth dependent and independent.

The term Y_{ATP} describes the theoretical yield in bacterial dry cells per mole ATP produced. Bauchop and Elsden (1960) originally suggested that Y_{ATP} was relatively constant and proposed a value of 10.5. Hespell and Bryant (1979) have suggested that Y_{ATP} could approach a maximum of 26 for a mixed rumen microbial population at an infinite growth rate. Since some of the ATP is used for maintenance, observed yields have been substantially less than this maximum and quite variable (Hespell and Bryant, 1979; Stern and Hoover, 1979; Van Soest, 1982). This can be attributed to the high cost of maintenance, especially at low growth rates. The high cost of maintenance may be due partly to energetic uncoupling that can be influenced by nutrient imbalances and by environmental factors such as ionic concentration and H + concentration. In a dynamic ecosystem, nutrients, such as branched chain VFA, ammonia, and amino acids, might be limiting at certain times after feeding.

Microbial efficiency as expressed in chemostat studies must be used in *in vivo* experiments with caution. Efficiency as reviewed by Bergen et al. (1982) is a rate of yield per unit of substrate in the rumen. The extent of substrate disappearance is coupled with the efficiency of yield and results in the microbial mass in the rumen at any time, which is yield per substrate. The rate of growth for any bacterial niche is a function of balanced substrate and nutrient availability per unit time. The pool size of microbial mass in the rumen is modified by liquid and solid outflow and protozoa predation. Animal measurements provide estimates of the flow of microbial matter to the abomasum or duodenum. The data are expressed as yield per substrate apparently or truly fermented. This is not a measurement of true efficiency, rather it is a measure of microbial wash-out and the amount of microbial matter recycled in the rumen. These interactions can be described by Michaelis-Menten kinetics (Bergen et al., 1982; Van Soest, 1982). Oldham (1984) suggested that microbial efficiency can be estimated by the equations originally derived by Pirt (1965) and summarized by Bergen et al. (1982). He further suggested that the microbial outflow be divided between that flowing with solids and that with liquid. Minato et al. (1966) presented and Oldham (1984) reviewed evidence that a significant proportion of the microbial population is associated with the particulate matter leaving the rumen. Oldham (1984) proposed the following equation:

$$K_m = P_s K_s + P_1 K_1,$$

where P_s and P_1 are the proportions of microbial population associated with the solid and liquid fractions, respectively, and K_s and K_1 are the fractional outflow rates for solids and liquids, respectively. This concept provides a biological and dynamic basis for predicting microbial flow. Unfortunately, the *in vivo* studies measuring rumen microbial efficiency are minimal and the predictability of flow of liquid and solids is relatively low (Evans, 1981a,b).

Forage intake has been shown to improve microbial flow (summarized by Johnson and Bergen, 1982; Van Soest, 1982). This may be caused by the combination of increased saliva flow, increased liquid turnover (increased small particle washout with attached bacteria), and increased pH, which could improve the ruminal environment, reduced total ruminal maintenance cost (older microbes being washed out), and a more juvenile population where the maintenance requirement is a small proportion of total requirement at high growth rates (Russell and Hespell, 1981).

Rumen dilution rate has been shown to have a significant impact on microbial flow (Ibrahim and Ingalls, 1971; Harrison et al., 1976; Kennedy et al., 1976; Kennedy and Milligan, 1978; Hartnell and Satter, 1979; Rogers et al., 1979; Bergen et al., 1982; Van Soest et al., 1982). Data summarized by Van Soest et al. (1982) employing the Michaelis-Menten relationship with *in vivo* and *in vitro* data give the average equation: $1/y = 0.14 + 0.015(1/x)$, $(R^2 = 0.76)$. In this equation y = g rumen microbial N/100 g organic matter fermented in the rumen adjusted for microbial incorporation of nutrient organic matter and x = fractional rumen liquid dilution rate. This equation was derived in part from steady-state data with animals at maintenance and resembles chemostat data (Van Nevel and Demeyer, 1976). Extrapolation of this equation beyond these data is not advised as these conditions need further verification.

Leng and Nolan (1984) have suggested that 30 to 50 percent of the total flux of ammonia was recycled through pathways within the rumen (ammonia $\longrightarrow$ other nitrogenous compounds $\longrightarrow$ ammonia). The nitrogen can come from lysed bacteria due to activity of bacteriophages and mycoplasmas and cell death. The latter can occur by starvation, especially under maintenance-fed or meal-fed conditions.

A significant amount of recycling can occur through protozoal predation of bacteria (Coleman, 1975). Generally, there is an inverse relationship between ruminal protozoa and bacteria concentrations. Coleman (1975), based on *in vitro* studies, suggests that more than 10^8 bacteria can be ingested per hour by the protozoal mass. Leng and Nolan (1984) feel that this is probably excessive.

It is suggested that recycling of bacterial N will be higher in conditions of lower intake where forage makes up a significant part of the diet. Also, recycling would be significant when animals are consuming diets multiple times per day or other conditions contributing to lower turnover rates, reduced washout of particles and microbial mass. More work is definitely needed in this area. Recycling of bacterial nitrogen must be taken into account in any estimate of microbial flow.

In order to predict microbial flow on efficiency it is necessary to know the amount of organic matter fermented. Johnson and Bergen (1982) have reviewed some of the recent literature. Their summary would suggest that processing, feed type, intake level, amount of forage consumed, and animal type may affect the extent of organic matter fermentation in the rumen.

Microbial flow has been determined in many experiments in sheep, beef cattle, and dairy cattle. These data are shown in Appendix Table 3 and are, in part, from the summary of Johnson and Bergen (1982). Extensive measurements have been made with sheep at or near maintenance. Fewer measurements have been made with beef cattle. These measurements have been observed under a broad range of feeding conditions and processing methods (Johnson and Bergen, 1982). Dairy cattle data are limited in number and source but relatively high intakes have been achieved.

Data are presented in Appendix Table 3 and regression summaries in Table 8. Estimates of TDN were made based on chemical analyses and ingredient composition (NRC, 1982). Diet DE (Mcal/kg DM) = 0.04409 TDN according to NRC (1982). Flow of microbial protein for the combined data set is correlated with dry matter intake ($r^2 = 0.65$). Slopes are similar for sheep and dairy cattle, but not beef cattle.

Dry matter intake will influence not only the quantity and possibly the type of substrate available for synthesis of microbial protein, but also various ruminal parameters such as pH and dilution rate and microbial determinants such as bacterial dilution rate, protozoal presence and bacterial numbers, distribution, and lysis in the rumen. These factors should be studied independently so that individual components can be used in predictive equations. Unfortunately, these factors cannot be eval-

TABLE 8   Regressions for Dairy Cattle, Sheep, and Beef Cattle

Model	Regressions	R^2	B_0	S.E.	B_1	S.E.	B_2	S.E.
Dairy regressions								
BY (gN/d)	DMI (kg/d)	.735	− 33.84	12.00	17.62	.98		
	OMI (kg/d)	.739	− 28.49	11.60	18.56	1.02		
	OMI less EE (kg/d)	.754	− 34.14	14.00	19.66	1.19		
	OMI less EE & Lignin (kg/d)	.737	− 51.09	18.87	24.24	1.82		
	DEI (Mcal/d)	.774	− 31.86	10.74	5.92	.29		
	TDNI (kg/d)	.774	− 31.86	10.74	26.12	1.30		
	adj. DEI (Mcal/d)	.762	− 47.14	14.04	6.68	.38		
	adj. TDNI (kg/d)	.762	− 47.14	14.04	29.47	1.71		
	RDOM	.627	− 14.23	16.38	29.05	2.27		
	TDOM	.726	− 24.21	15.99	25.55	1.75		
	FI & CI	.742	− 34.57	11.91	16.11	1.29	19.54	1.47
BY/RDOM (gN/d/kg)	DMI	.212	16.64	2.03	.812	.160		
	OMI	.211	16.87	1.99	.854	.168		
	DMI less EE	.250	15.86	2.22	.966	.186		
	OMI less EE & Lignin	.298	14.10	2.85	1.32	.269		
	DEI (Mcal/d)	.242	16.18	1.95	28.82	5.22		
	TDNI (kg/d)	.242	16.18	1.95	1.27	.23		
	adj. DEI	.267	14.39	2.27	34.49	6.20		
	adj. TDNI	.267	14.39	2.27	1.52	.27		
BY/adj. DEI (gN/d/Mcal)	DMI (kg/d)	.127	361.52	42.56	11.94	3.26		
	OMI (kg/d)	.134	360.68	41.68	12.93	3.43		
Sheep regressions								
BY (gN/d)	DMI (kg/d)	.685	1.61	.84	11.81	.98		
	OMI (kg/d)	.693	1.27	.85	13.18	1.06		
	OMI less EE (kg/d)	.082	5.46	2.32	6.10	3.49		
	OMI less EE & Lignin (kg/d)	.150	3.76	2.36	9.92	4.04		
	DEI (Mcal/d)	.729	− 1.29	.96	5.22	.39		
	TDNI (kg/d)	.729	− 1.29	.96	23.04	1.71		
	adj. DEI (Mcal/d)	.767	− 2.14	.94	5.50	.38		
	adj. TDNI (kg/d)	.767	− 2.14	.94	24.26	1.68		
	RDOM	.548	− 2.01	1.49	27.57	3.04		
	TDOM	.644	− 3.21	1.49	24.45	2.48		
	FI & CI	.729	− .37	.99	13.08	.99	17.32	1.91
BY/RDOM (gN/d/kg)	DMI	.142	17.39	1.87	7.27	2.18		
	OMI	.142	17.23	1.91	8.06	2.40		
	DMI less EE	.108	31.34	4.85	− 14.80	7.29		
	OMI less EE & Lignin	.082	30.58	5.19	− 15.53	8.90		
	DEI (Mcal/d)	.114	16.61	2.35	2.79	.95		
	TDNI (kg/d)	.114	16.61	2.35	12.31	4.19		
	adj. DEI	.131	15.82	2.29	2.28	.93		
	adj. TDNI	.131	15.82	2.29	12.72	4.12		
BY/adj. DEI (gN/d/Mcal)	DMI (kg/d)	.023	418.13	31.47	44.78	36.50		
	OMI (kg/d)	.025	416.08	32.54	50.89	40.38		
Beef regressions								
BY (gN/d)	DMI (kg/d)	.470	13.74	3.84	6.24	.78		
	OMI (kg/d)	.533	7.14	4.22	8.40	.87		
	OMI less EE (kg/d)	.445	10.35	4.82	7.70	1.14		
	OMI less EE & Lignin (kg/d)	.260	18.42	5.44	6.56	1.47		
	DEI (Mcal/d)	.322	16.79	5.06	1.81	.34		
	TDNI (kg/d)	.332	16.79	5.06	7.97	1.49		
	adj. DEI (Mcal/d)	.226	17.74	6.10	1.76	.43		
	adj. TDNI (kg/d)	.226	17.74	6.10	7.77	1.91		
	RDOM	.238	26.29	4.21	7.42	1.49		
	TDOM	.281	18.24	5.34	7.60	1.69		
	FI & CI	.490	12.88	4.17	7.36	1.10	5.89	.86

TABLE 8 *Continued*

Model	Regressions	R^2	B_0	S.E.	B_1	S.E.	B_2	S.E.
BY/RDOM								
(gN/d/kg)	DMI	.027	22.12	2.70	− .80	.57		
	OMI	.011	20.65	2.44	− .48	.52		
	DMI less EE	.018	20.78	2.91	− .69	.686		
	OMI less EE & Lignin	.078	23.79	2.75	− 1.63	.741		
	DEI (Mcal/d)	.114	25.52	2.88	− .54	.20		
	TDNI (kg/d)	.114	25.52	2.88	− 2.39	.884		
	adj. DEI	.123	26.05	2.94	− .59	.21		
	adj. TDNI	.123	26.05	2.94	− 2.60	.92		
BY/adj. DEI								
(gN/d/Mcal)	DMI (kg/d)	.0064	341.80	48.85	− 6.33	10.44		
	OMI (kg/d)	.0032	333.56	49.49	− 4.84	11.35		

DMI = Dry matter intake (kg/d)
OMI = Organic matter intake (kg/d)
OMI, less EE = Organic matter intake, corrected for ether extract (kg/d)
OMI, less EE & Lignin = Organic matter intake, corrected for ether extract and lignin (kg/d)
DEI = Digestible energy intake (Mcal/d)
TDNI = Total digestible nutrients intake (kg/d)
adj. DEI = Digestible energy intake, adjusted for maintenance (Mcal/d)
adj. TDNI = Total digestible nutrients intake, adjusted for maintenance (kg/d)
RDOM = Rumen digested organic matter (kg/d)
TDOM = Total tract digested organic matter (kg/d)
FI = Forage intake (kg/d)
CI = Concentrate intake (kg/d)
BY = Bacterial yield (gN/d)

uated independently using the currently available data base concerning microbial protein synthesis. Indeed, the addition of digestibility of the diet to the regression equations developed did not improve the ability to predict microbial yield. Further, the accuracy of predicting microbial efficiency (BY/ROMD) was very poor. Further studies are needed to evaluate these factors independently so that diet digestibility, extent of digestion in the rumen and the effect of dilution rate, protozoal presence, and pH on microbial efficiency can be determined and employed to improve efficiency of synthesis of microbial protein in the rumen.

The amount of organic matter fermented in the rumen is dependent on the rate of digestion and the rate of passage (Mertens and Ely, 1979; Van Soest et al., 1979; 1982). This would follow the equation

$$D = \frac{K_d}{K_d + K_p},$$

where D = organic matter digestion, K_p = rate of passage, and K_s = rate of digestion.

As organic matter passage increases, the amount of potentially fermentable organic matter actually fermented is reduced unless K_d is very high relative to K_p. External phenomena not associated with feed type and rate of passage may alter this and, along with such events as a delay in microbial attachment to substrate, might delay digestion. Van Soest et al. (1982) have suggested that a time lag in digestion (due to hydration or other phenomena) also may delay passage.

Johnson and Bergen (1982) adjusted the estimates of ruminally fermented organic matter for microbial mass, thus giving an estimate of truly fermented organic matter, and calculated percentage of total tract digestion occurring in the rumen. For cattle trials for which total tract digestibilities were available, percent of total tract digestion occurring in the rumen was 76 ± 10. They expressed the true ruminally fermented organic matter as a percent of total tract organic matter digestion. The variability of digestion in the rumen was increased, and it was suggested that this was due in large part to differences in microbial yields. The largest data base is with animals fed at maintenance, and thus a true evaluation of the impact of the variation is not possible.

Measurement techniques for site of digestion studies are responsible for some of the variation. Solid phase digesta markers, which are not well attached, can migrate from the treated particle to other particles, or move with the liquid phase. Microbial markers are a source of variance. While RNA marks bacteria and protozoa, diamino-pimelic acid (DAP) identifies only bacteria. However, all of these systems suffer from various limitations. Nevertheless, these are the systems upon which most of the data are founded.

TABLE 9   Empirical Prediction Models of Microbial Nitrogen Flow

Model	R^2	B_0	S.E.	B_1	S.E.	B_2	S.E.	B_3	S.E.	B_4	S.E.	B_5	S.E.	B_6	S.E.
BY (gN/d)															
= BW + DMI + F% + TDN + RDOMP	.939	− 114.60	52.36	− .031	.029	18.48	.92	.293	.157	1.23	.683	− .061	.309		
= BW + OMI + F% + TDN + RDOMP	.940	− 108.70	51.58	− .037	.029	19.93	.97	.280	.155	1.14	.67	− .030	.304		
= BW + FORGEI + CONCEI + TDN + RDOMP	.938	− 5.65	39.37	− .058	.028	17.85	1.22	21.04	1.84	− .265	.625	.127	.32		
= BW + F% + RDP + OME + TDN + RDOMP	.949	− 131.31	48.53	− .026	.027	.198	.146	9.23	2.88	33.46	3.58	.505	.650	1.11	.407
= BW + F% + RDOM + OME + TDOM + RDOMP	.948	− 96.23	32.04	− .031	.027	.111	.098	8.87	2.87	34.17	3.47	− .054	.340	1.25	.368
= BW + F% + TDNI + RDOMP	.942	− 49.26	20.32	− .030	.027	.347	.099	27.88	1.29	.182	.282				
= BW + F% + TDNAI + RDOMP	.942	− 55.22	20.29	− .06	.028	.356	.099	31.92	1.48	.244	.281				
= BW + F% + TDN + PDOMI + RDOMP	.922	− 64.37	59.61	.011	.032	.347	.177	.312	.780	23.07	1.34	− .012	.349		
= BW + F% + TDNIM + NEEOMI	.940	− 113.85	51.31	− .036	.029	.284	.153	1.19	.633	20.62	.994				
= BW + F%	.671	− 75.88	18.20	.464	.032	.638	.227								
= BW + F% + DEAI	.89	− 28.1	—	− .074	.016	.29	.07	6.75	.24						
= MBW + F% + DEAI	.89	− 22.0	—	− .44	.09	.28	.07	6.83	.24						
= DEIBW + F%	.33	− 70.2	—	.26	.02	.128	.17								
= DEIMW + F%	.74	− 79.8	—	.07	.003	.31	.11								
= DEAIBW + F%	.27	− 64.2	—	25.9	2.91	.13	.18								
= DEAIMW + F%	.72	− 112.0	—	8.14	.35	.35	.11								
= BW + DEI + F%	.87	− 24.6	—	− .054	.02	6.0	.2	.26	.069						
= MBW + DEI + F%	.90	− 20.0	—	− .32	.08	6.1	.2	.26	.069						
= BW + DEI	.90	− 10.1	—	− .039	.02	5.88	.28								
= BW + DEAI	.90	− 11.9	—	− .060	.02	6.57	.32								
= DEIBW	.37	− 70.6	—	28.3	3.5										
= DEAIBW	.31	− 66.6	—	28.3	4.1										
= MBW + DEI	.91	− 6.49	—	− .25	.11	5.96	.27								
= MBW + DEAI	.90	− 6.99	—	− .36	.11	6.67	.32								
= DEIMW	.76	− 79.2	—	7.44	.4										
= DEAIMW	.74	− 89.8	—	8.14	.46										

BY = Bacterial yield (gN/d)
BW = Body weight (kg)
DMI = Dry Matter Intake (kg/d)
OMI = Organic Matter Intake (kg/d)
F% = Percent forages in ration
TDN = % total digestible nutrients in diet (%)
RDOMP = % rumen digested organic matter (% of organic matter intake)
FORGEI = Forage intake (kg/d)
CONCEI = Concentrate intake (kg/d)
OME = Organic matter escaped (kg/d)

TDNI = Total digestible nutrients (kg/d)
PDOMI = Organic matter intake, corrected for ether extract (kg/d) + lignin × 2.4 (kg/d)
NEEOMI = Organic matter intake, corrected for ether extract and lignin (kg/d)
MBW = $BW^{.75}$
DEI = Digestible energy intake (Mcal/day)
DEIBW = Digestible energy intake (Mcal/100 kg BW/day)
DEIMW = Digestible energy intake (Mcal/100 kg $BW^{.75}$/day)
DEAI = DE intake depressed 4% per multiple intake (Mcal/day)
DEAIBW = DE intake depressed 4% per multiple intake (Mcal/100 kg BW/day)
DEAIMW = DE intake depressed 4% per multiple intake (Mcal/100 kg $BW^{.75}$/day)

Validity of various microbial markers also has been questioned. Various workers have reviewed and compared these techniques (Smith, 1979; Stern and Hoover, 1979; Steinhour and Clark, 1982). Several workers have compared microbial markers (Harmeyer et al., 1976; Ling and Buttery, 1977; Siddons et al., 1979; Wolstrup et al., 1979; McAllan and Smith, 1983). McAllan and Smith (1983) recently demonstrated little difference between RNA and DNA microbial concentrations in microbes from defaunated sheep. Bergen et al. (1982) have shown that the RNA/protein ratio increases with increased microbial growth rate, suggesting that a careful definition of the physiological state of the microorganisms being sampled is needed in order to interpret the information. Work is still needed in this area, especially to identify a primary standard from which predictive methods can be developed. Mason (1969) provides evidence that it might be more appropriate to estimate microbial mass by difference through a combination of detergents and centrifugation. This approach needs more study.

Several multiple regression analyses were also performed using both linear and quadratic models (Table 9). Rumen models (Nolan et al., 1972; Baldwin and Denham, 1979; Black et al., 1980–1981) have integrated the current knowledge of the biology of microbial growth in the rumen. Some of the models suggest more than one microbial pool based on either substrate affinities or niches. Unfortunately, these models, although providing us with an improved understanding of the mechanisms of microbial growth and flow, are not advanced enough to use in a quantitative field application model. An alternative is the use of empirical models. These are presented in Table 9. The simplest model that may have field application is one in which the parameter(s) can be readily measured in the field such as dry matter or organic matter intake.

The regressions for the combined data are presented in Table 9. The models demonstrate the importance of the interactions between forage, energy value of the diet, and intake. Further research is needed to develop quantitative dynamic prediction models that will incorporate measurements of diet type and processing (Johnson and Bergen, 1982), rumen escape estimates, and potential substrate degradability on various microbial niches in the rumen and flow of microbes from the rumen. Further, it is important that studies be conducted to determine the interactions of N recycling in the rumen and its importance on microbial flow to the small intestine.

The interactions of intake, diet type, and rumen volume with microbial efficiency in the rumen and microbial flow to the small intestine are complex. The present data set does not adequately allow the development of one equation that will describe these interactions for dairy cattle, beef cattle, and sheep. Separate equations are therefore recommended for each species and are summarized in Table 10. The equations for dairy cattle and sheep adequately describe microbial yield based on TDN intake. The beef equation is for rations containing less than 40 percent of forage (see Table 10) and includes forage and concentrate components. The use of TDN is suggested because it represents the largest data base that is available on feedstuffs today, and the vast majority of feed analysis laboratories base the predicted energy content of feeds on an equation driven at some point by TDN. TDN is a good estimate of whole tract DOM. The equations for NEL are derived directly from TDN (NRC, 1978) and are presented for convenience.

TABLE 10 Equations Used for Predicting Microbial Yield or Efficiency

Item	Dairy Cattle	Sheep	Beef Cattle[a]
Dependent variables	Microbial N, g/d	—	(Microbial N, g/TDNI, kg)
Independent variables			
Intercept	-31.9 (10.7)[b]	-1.29 (0.96)	8.63 (1.67)
TDN intake, kg	26.13 (1.3)	23.0 (1.71)	—
Forage intake, kg	—	—	14.6 (2.84)
(Forage intake, kg)^2	—	—	-5.18 (1.3)
Concentrate intake	—	—	0.595 (0.8)
r^2	0.77	0.73	0.36[c]
Independent variables			
Intercept	-30.92 (10.69)	—	—
NEL, Mcal/d	11.45 (0.57)	—	—
r^2	0.77	—	—

[a] Microbial yield, gN/day = TDNI × Microbial N, g/TDNI. To be used for cattle receiving less than 40 percent of their DMI as forage.

[b] Standard error.

[c] The use of this equation improves the prediction (r^2.0.58) of microbial flow compared to the use of TDN intake alone.

Digestion and Absorption in the Small Intestine

NITROGEN SUPPLY

Because of N transactions in the rumen, intake of protein (IP) is not an accurate indicator of N flow to the small intestine (Whitlow and Satter, 1979). Examination of data obtained with cows, steers, calves, sheep, and lambs showed that for both concentrate and forage diets, duodenal flow (SCP) ranged between 10.5 and 12.5 g NAN (nonammonia nitrogen) per Mcal ME consumed (Oldham and Tamminga, 1980). This illustrates that energy consumption is the major determinant of the amount of N entering the small intestine. As discussed in other sections, factors that affect microbial protein production and ruminal degradation of dietary protein can be expected to modify N supply (SCP) to the small intestine.

The N entering the duodenum is a combination of microbial (BCP), undegraded intake (UIP) and endogenous protein. Differences in amino acid composition between bacteria and protozoa (Chalupa, 1972), IP and BCP (Smith, 1975; Laughren and Young, 1979), and IP and UIP (Smith and Mohamed, 1977) imply that the quantities of BCP and UIP entering the intestine can influence the supply of absorbable amino acids. Unfortunately, technical problems in partitioning BCP and UIP and in estimating endogenous secretions have made accurate quantitation difficult. The N entering the duodenum from the stomach can range between 0.3 to 1.0 BCP and 0 to 0.70 UIP (Smith, 1975). Endogenous influxes can equal the supply from the stomach (Nolan, 1975)

From a nutritional point of view, it is important to know the chemical composition of intestinal N. The proportionate distribution of N in duodenal contents of cows (SCP) was described by Oldham and Tamminga (1980) as essential amino acids, 0.35; nonessential amino acids, 0.30; amides, 0.04; nucleic acids, 0.11; ammonia, 0.06; and unknown, 0.14. Increases in the ratio of essential amino acids to nonessential amino acids in duodenal digesta as a function of the concentration of intake protein (IPDM) indicate that the amino acid composition of IP may influence the balance of amino acids available for absorption (Laughren and Young, 1979). Because it is measured easily, NAN is a frequently used measure of N entering the small intestine. In data summarized by Stern and Satter (1982), amino acid N in the duodenum of lactating cows was 0.79 of NAN. In sheep, amino acid N was 0.80 of NAN (Hogan and Weston, 1970). Considerable work is needed before it will be possible to predict the amounts of specific amino acids presented to the small intestine.

DIGESTION SYSTEM

Digestion of protein in the abomasum and small intestine appears to be the same for ruminants as in nonruminants except for the slow neutralization of digesta in the small intestine and the abundance of pancreatic ribonuclease (Armstrong and Hutton, 1975; Bergen, 1978; Chalupa, 1978).

Slow neutralization of digesta in the upper small intestine of ruminants appears to be related to the low bicarbonate content of pancreatic juice (Taylor, 1962). This extends the activity of abomasal pepsin but delays the onset of activities of intestinal enzymes. Thus, considerable proteolysis in the duodenum is due to the gastric protease, pepsin. Optimal activity for trypsin, chymotrypsin, and carboxypeptidase does not occur until the middle jejunum, and peak activity of exopeptidases and dipeptidases is found in the mid ileum (Ben-Ghedalia et al., 1974).

Breakdown of nucleic acids is achieved by DNases,

RNases, phosphodiesterases, and phosphomonesterases (Bergen, 1978; Roth and Kirchgessner, 1980). An important role for abundant pancreatic RNase in the ruminant is release of nucleic acid phosphorus for recycling to the rumen via saliva (Barnard, 1969). It appears that the products of nucleic acid digestion that are absorbed are nucleotides, nucleosides, and bases (Bergen, 1978; Smith, 1979).

ABSORPTION MECHANISMS

The mucosa of the small intestine contains uptake systems for free amino acids, peptides, nucleotides, and nucleosides (Armstrong and Hutton, 1975; Bergen, 1978; Scharrer and Amann, 1980).

The most active site for amino acid absorption in sheep is the mid to lower ileum (Johns and Bergen, 1973; Phillips et al., 1976), but the highest rate of amino acid disappearance *in situ* from the digesta in the small intestine occurs in the mid jejunum (Ben-Ghedalia et al., 1974). Johns and Bergen (1973), using jejunal strips, demonstrated that amino acid uptake in sheep occurs against a concentration gradient, exhibits saturation kinetics, and depends upon metabolic energy. K_m and V_{max} values for glycine, methionine, and lysine transport in sheep jejunum were similar to values obtained with rat jejunum. The preferential disappearance of essential amino acids over nonessential amino acids from digesta flowing through the small intestine has been demonstrated in sheep (Johns and Bergen, 1973; Phillips et al., 1976) and in cattle (Van't Klooster and Boekholt, 1972). Using exteriorized intestinal loops, Williams (1969) ranked amino acid absorption as follows:

Ile > Arg ≥ Val > Leu > Met > Phe > Lys > Try >
    Asp ≥ Ser > Ala > Pro > His ≥ Thr ≥ Glu > Gly.

The order with jejunal strips *in vitro* was Met > Lys > Gly (Johns and Bergen, 1973) and with everted sacs *in vitro* was Met > Val > Thr (Phillips et al., 1976). The overall order of uptake by sheep gut is similar to that noted in man and rats. A depressing effect of leucine on lysine uptake has been shown both *in vitro* (Johns and Bergen, 1973) and *in vivo* (Hume et al., 1972).

It seems likely that as in the nonruminant (Matthews, 1972; Munck, 1976), absorption of peptides is quantitatively important in the ruminant. Steps involved include peptide uptake, peptide hydrolysis, and transport of amino acids.

Removal of the end products of nucleic acid digestion from digesta flowing through the small intestine implies efficient absorption mechanisms (Bergen, 1978). Nucleosides are absorbed from the small intestine by a Na-dependent saturatable transport process (Scharrer and Amann, 1980).

EXTENT OF APPARENT ABSORPTION

Measuring disappearance of N (SCP) or amino (STP) acids between the duodenum and ileum provides an estimate of apparent absorption. Samples from cannulae inserted into the duodenum prior to the entry of bile and pancreatic secretions only includes endogenous N from gastric secretions, whereas samples from cannulae inserted posterior to the entry of bile and pancreatic secretions also contain N from pancreatic secretions.

Apparent absorption of NAN and amino acids from the small intestine of lactating cattle, nonlactating cattle and sheep fed a variety of diets is listed in Appendix Tables 11, 12, and 13. Table 11 summarizes the results. Overall, apparent absorption was 0.65 of NAN and 0.68 of amino acids entering the duodenum (Table 11). Apparent absorption of NAN was similar in the groups summarized and was less than absorption of amino acids in lactating cattle and sheep but not in nonlactating cattle. In experiments reviewed by ARC (1980) in which absorption of both NAN and amino acids were measured, the values did not differ markedly. However, Tamminga (1980) concluded that apparent absorption of total N is usually 0.05 lower than that of amino acids.

Based upon the foregoing, values suggested for apparent absorption of NAN and amino acids from the small intestine are 0.65 and 0.70 of amounts entering the duodenum.

Apparent absorption of essential amino acids is about 0.05 greater than nonessential amino acids (Armstrong et al., 1977; Tamminga, 1980). Apparent absorption of essential amino acids, as summarized by Tamminga (1980) suggests that absorption of lysine and arginine is greater while absorption of threonine, valine, and phenylalanine is less than the absorption of total essential amino acids (Table 12). Apparent absorption of

TABLE 11   Summary of Apparent Absorption of Nonammonia Nitrogen and Amino Acids from the Small Intestine of Ruminants[a]

Measurement	Calculation			
	n^b	$\bar{x}$	SD	CV
NAN				
Lactating cattle	12	0.65	0.04	0.07
Nonlactating cattle	17	0.66	0.04	0.06
Sheep	29	0.64	0.06	0.09
All	58	0.65	0.05	0.08
Amino acids				
Lactating cattle	21	0.69	0.05	0.08
Nonlactating cattle	11	0.62	0.06	0.10
Sheep	22	0.70	0.06	0.09
All	54	0.68	0.06	0.10

[a]Based upon data in Appendix Tables 11, 12, 13.
[b]Number of diets.

TABLE 12   Proportionate Disappearance of Amino Acids from the Small Intestine[a]

Amino Acid	Sheep	Cow I[b]	Cow II[b]	Cow III[b]
		Animal and Experiment		
Lysine	0.77	0.77	0.74	0.75
Histidine	0.70	0.77	0.73	0.76
Arginine	0.79	0.82	0.79	0.79
Threonine	0.68	0.70	0.71	0.71
Valine	0.72	0.72	0.72	0.72
Methionine	0.75	0.75	0.61	0.66
Isoleucine	0.73	0.76	0.72	0.72
Leucine	0.74	0.78	0.73	0.73
Phenylalanine	0.69	0.68	0.71	0.72
Total essential amino acids	0.73 ± 0.008	0.75 ± 0.005	0.72 ± 0.012	0.74 ± 0.005
No. observations	21	8	13	72

[a]Data summarized by Tamminga (1980).

[b]Cow I, II, and III are not individual cows but refer to experiments involving cows.

methionine was quite variable in the cow experiments and may be a consequence of location of the duodenal cannula in that only absorption of methionine was lower in experiments where duodenal samples were collected beyond rather than prior to the pancreatic and biliary duct. In experiments reviewed by Armstrong et al. (1977), apparent absorption of methionine was 0.06 ± 0.05 more than apparent absorption of total essential amino acids. Net disappearance of cystine was only 0.40 to 0.50.

ENDOGENOUS LOSS

Calculation of true absorption requires correction for the influx of endogenous nitrogen that is not reabsorbed from the small intestine.

Endogenous protein enters the small intestine in the form of enzymes, bile, mucus, serum albumin, lymph, epithelial cells, and other degradable products from the gastrointestinal lining (Swanson, 1982). Summation of the endogenous input to the entire gastrointestinal tract is large (Phillipson, 1964; Swanson, 1982). In nonlactating cattle, it is more than twice the maintenance value (Swanson, 1982). In studies where ^{15}N was used to study N metabolism in sheep, inputs (g/d) of NAN to the small intestine were: UIP, 6.5; BCP, 10.3; and intestinal secretions, 17.0 (Nolan, 1975). Thus, endogenous N was equivalent to NAN from the stomach. The N in duodenal contents from abomasal juice, pancreatic juice, bile, and epithelial cells was estimated at 0.004 dry matter consumed (Tamminga et al., 1979). In lactating cattle, this was equivalent to 0.10 to 0.15 of N from the stomach, but not included in this estimate are endogenous inputs posterior to the entry of bile and pancreatic juice.

Total influx of endogenous N is important for an understanding of the dynamic involvement of intestinal tissue in N metabolism. However, as indicated previously, calculation of the true intestinal absorption of N derived from the stomach requires correction for the endogenous input that is not reabsorbed.

Endogenous losses, as well as true digestibility, can be estimated as $Y = a + bX$ where Y is disappearance between two points in the gastrointestinal tract (i.e., mouth and anus; proximal duodenum and terminal ileum), X is the supply (amount or concentration) to a point in the gastrointestinal tract (i.e., mouth; proximal duodenum), a is a negative value and represents the endogenous loss, and b is true digestibility (Van Soest, 1982).

Applying the regression approach to data obtained with sheep, Hogan and Weston (1970) calculated the endogenous loss from the small intestine that appeared in feces as 0.0016 organic matter entering the duodenum. The endogenous loss appearing in feces from the entire gastrointestinal tract was 0.004 organic matter consumed. This prompted Hogan and Weston (1970) and Hogan (1975) to conclude that only about one-third of the N in the classic metabolic fecal fraction is of endogenous origin and the remaining two-thirds is of microbial origin.

Regression analysis of other data yielded the following estimates of endogenous losses from the small intestine (g/day): sheep, 2.2 g NAN; lactating cattle, 56 g NAN and 250 g amino acids; nonlactating cattle, 0.77 g NAN and 98 g amino acids (Table 13). In sheep, Tas et al. (1981) estimated the mean endogenous loss of amino acids secreted into the small intestine to be 13 g/d. Since NAN is 0.8 amino acid nitrogen, this is equivalent to 2.6 g NAN/d [13 ÷ (0.8 × 6.25)].

The endogenous losses in Table 13 for sheep and lactating cattle are 0.10 to 0.13 of the N supply to the proximal duodenum. With nonlactating cattle, the endogenous loss of amino acids was equivalent to 0.16 of the supply from the stomach. The endogenous loss of NAN, however, was only 0.01 of duodenal NAN. As shown in Table 13, estimates for the two data sets were not in agreement (Zinn and Owens, 1982, 0.01; Sharma et al., 1974, 0.24).

EXTENT OF TRUE ABSORPTION

NAN and Amino Acids from the Stomach

Regression analysis used to estimate endogenous losses from the small intestine also provided estimates of true absorption (Table 13). Expressed as a proportion of the N supply to the proximal duodenum, values obtained were: sheep, 0.75 NAN; lactating cattle, 0.78 NAN and

TABLE 13 Supplies of Nonammonia Nitrogen and Amino Acids to the Proximal Duodenum and Terminal Ileum, Apparent and True Digestibility of Nitrogen, and Endogenous Loss of Nitrogen to the Small Intestine

Source of Data	No.[a]	Nitrogen Fraction	Supply to Proximal Duodenum (g/d)	Supply to Distal Ileum (g/d)	Apparent Disappearance from Small Intestine (g/d)	Apparent Absorption	True Absorption	Endogenous Loss[d] (g/d)	Proportion of Duodenal	Correlation (R^2) of Disappearance (Y) and Supply (X)
Sheep										
Lu et al. (1981)										
Merchen and Satter (1983a)	27	NAN[b]	22.6	8.0	14.6	0.64	0.75**	2.2**	0.10	0.96
MacRae et al. (1972)			±1.0[c]	±1.6[c]	±0.8[c]	±0.04[c]	±0.03[e]	±0.8		
Ørskov et al. (1971a,b; 1972, 1974)										
Nonlactating cattle										
Zinn and Owens (1982)	13	NAN	102.9	33.8	69.1	0.67	0.68**	1.2	0.01	0.97
			±8.9	±3.0	±6.2	±0.01	±0.04	±3.9		
Sharma et al. (1974)	4	NAN	124.5	45.6	78.9	0.63	0.87*	29.5	0.24	0.89
			±7.1	±2.3	±6.5	±0.02	±0.21	±25.9		
Combined	17	NAN	108.0	36.6	71.4	0.66	0.67**	0.77	0.01	0.95
			±7.3	±2.6	±5.0	±0.01	±0.04	±4.3		
Sharma et al. (1974)	4	Amino acids	615.9	202.3	413.6	0.67	0.83*	97.8	0.16	0.93
			±44.1	±12.3	±37.9	±0.02	±0.16	±97.6		
Lactating cows										
Merchen et al. (1983b)	12	NAN	442.6	155.0	287	0.65	0.78**	56.0	0.13	0.88
Merchen (1981)			±19.3	±7.1	±16.0	±0.01	±0.09	±40.9		
Stern and Satter (1982)										
Stern et al. (1984)	17	Amino acids	2092	609	1455	0.70	0.82**	250	0.12	0.92
Pena et al. (1985)			±97	±45	±82	±0.01	±0.06	±135		

[a]Number of diets.
[b]Nonammonia nitrogen.
[c]Standard error of the mean.
[d]Estimated from $Y = a + bx$ where Y = apparent disappearance (g/d) from the small intestine, x = supply to proximal duodenum (g/d), a = endogenous input that is not reabsorbed, and b = true digestion.
[e]Standard error of estimate.

*Different from 0 at $P < 0.05$.
**Different from 0 at $P < 0.01$.

0.82 amino acids; nonlactating cattle, 0.67 NAN and 0.83 amino acids.

When endogenous loss is expressed as a proportion of the N supply to the duodenum, true absorption is the sum of apparent absorption and endogenous loss (i.e., apparent absorption = 0.65; endogenous loss = 0.10; then true absorption = 0.75). Thus, the low value for true absorption of duodenal NAN from the small intestine of nonlactating cattle is a consequence of the small correction for endogenous loss.

Our estimates of true absorption obtained by regression analysis are in agreement with the few available published reports (Table 14). With 23 diets, true absorption of NAN from the small intestine of sheep was 0.76; true absorption of essential amino acids was 0.80 (Hogan and Weston, 1970). Tas et al. (1981) obtained a mean value in sheep of 0.86 for true absorption of amino acids. Using data from Nolan's (1975) model of N utilization in sheep, true absorption was 0.80 of the NAN supply to the duodenum. Values suggested for true ab-

sorption of NAN and amino acids from the small intestine are 0.75 and 0.80 of amounts entering the duodenum.

Microbial Protein

Information on the true digestibility of microbial N (BCP) is summarized in Table 14. In experiments reviewed by Bergen (1978), digestion of pure cultures of rumen bacteria *in vitro* ranged from 0.44 to 0.93. In data summarized by Chalupa (1972) and Zinn and Owens (1982), true absorption of rumen bacterial and protozoal protein in rats was 0.66 and 0.88, respectively. Labeling ruminal bacteria with ^{35}S yielded values of 0.74 (Bird, 1972) and 0.85 (Salter and Smith, 1977). Early studies with ^{15}N-labeled rumen bacteria (Smith et al., 1974) gave low and variable estimates of true absorption (0.41 to 0.70). In later studies (Salter and Smith, 1977) a value of 0.79 was obtained. ^{15}N can label compounds such as DAP to give variable and low

TABLE 14   Summary of True Absorption From the Small Intestine

Fraction	Specie	Method	Source of Data	True Absorption
NAN	Sheep	Regression	Table 12	0.78
	Sheep	Regression	Hogan and Weston (1970)	0.76
	Sheep	^{15}N	Nolan (1975)	0.80
	Lactating cattle	Regression	Table 12	0.78
	Nonlactating cattle	Regression	Table 12	0.67
			$\bar{X}$	0.76
			SD	0.05
			CV(%)	0.06
Amino acids	Sheep (EAA)	Regression	Hogan and Weston (1970)	0.80
	Sheep	Regression	Tas et al. (1981)	0.86
	Lactating cattle	Regression	Table 12	0.82
	Nonlactating cattle	Regression	Table 12	0.83
			$\bar{X}$	0.83
			SD	0.03
			CV	0.03
Microbial N (BCP)	Rat	Isolated bacteria	Chalupa (1972)	0.66[a]
	Rat	Isolated protozoa	Zinn and Owens (1982)	0.88[a]
	Sheep	^{35}S-bacteria	Bird (1972)	0.74
	Sheep	^{35}S-bacteria	Salter and Smith (1977)	0.85
	Sheep	^{15}N-bacteria	Salter and Smith (1977)	0.79
	Nonlactating cattle	Regression	Zinn and Owens (1982)	0.73
			$\bar{X}$	0.78
			SD	0.05
			CV(%)	0.06
Microbial amino acids (BTP)	Sheep	Regression	Tas et al. (1981)	0.87
Escape N (UIP)	Sheep	^{15}N Leaf protein	Salter and Smith (1977)	0.85
	Sheep	^{14}C Chloroplast protein	Smith et al. (1974)	0.73–0.82
	Nonlactating cattle	Regression	Zinn and Owens (1982)	0.68
			$\bar{X}$	0.77
			SC	0.09
			CV(%)	0.11
Escape amino acids	Sheep	Regression	Tas et al. (1981)	0.82
Endogenous amino acids	Sheep	Regression	Tas et al. (1981)	0.78–0.84

[a] Assuming equal biomasses of bacterial and protozoal nitrogen.

estimates of true absorption. Multiple regression analysis yielded estimates of 0.87 for microbial amino acids (Tas et al., 1981) and 0.73 for microbial N (Zinn and Owens, 1982). However, in the data of Zinn and Owens (1982), estimates of endogenous losses were small and true absorption of NAN was almost identical to apparent absorption.

As indicated in a previous section, microbial N is 0.10 to 0.20 nucleic acid N. Data summarized by Bergen (1978) and Smith (1975) indicate that digestion and absorption of nucleic acids is an efficient process. In sheep and cattle, 0.75 to 0.90 of the nucleic acids that enter the proximal duodenum are removed prior to the terminal ileum.

Undegraded Intake Protein

Estimates of the true absorption of protein that escapes ruminal degradation (UIP) have been obtained by isotopically labeling plant materials and by regression analysis (Table 14). Absorption of leaf protein labeled with ^{15}N was 0.85 (Salter and Smith, 1977), while 0.73 to 0.82 of ^{14}C-labeled chloroplast protein was absorbed (Smith et al., 1974). By regression analysis, the true absorption of escape amino acids in sheep was 0.82 (Tas et al., 1981). The true absorption of escape NAN in non-lactating cattle was, however, only 0.68 (Zinn and Owens, 1982).

Endogenous Nitrogen

Information on the true absorption of amino acids in the endogenous influx to the small intestine is scarce. Applying regression techniques to data from sheep (Tas et al., 1981) yielded values of 0.78 to 0.84 (Table 14).

NITROGEN METABOLISM IN INTESTINAL TISSUE

A substantial part of most amino acids apparently absorbed from the small intestine are metabolized in processes associated with absorption (Bergman and Heitman, 1978; MacRae, 1978; Tamminga and Oldham, 1980).

An estimate of amino acids metabolized by intestinal tissue can be obtained by comparing amino acids disappearing from the small intestine with those appearing in portal blood. In sheep fed 800 g/d of a high-protein diet (19.8 percent) or 650 g/d of a medium-protein (15.6 percent) diet, 0.67 to 0.71 percent and 0.55 to 0.57 percent, respectively, of the amino acids absorbed from the small intestine were metabolized in the intestinal wall (Tagari and Bergman, 1978). No preference appeared for either essential or nonessential amino acids.

Measurements of absorption based upon appearance of amino acids in blood (Hume et al., 1972; Sniffen and Jacobson, 1975), therefore, reflect the balance of removal from intestinal contents and metabolism in intestinal tissue.

SYNOPSIS

Summaries of apparent and true absorption of NAN and amino acids from the small intestine of sheep and cattle are in Tables 11 and 13. These data suggest that absorption from the small intestine does not vary greatly.

True absorption of microbial N (BCP) and N from dietary protein (IUP) that escaped ruminal degradation are similar. This is expected on the basis of the constancy of apparent and true absorption of NAN and amino acids. Bacterial cells might be less digestible because of mucopeptides in bacterial cell walls. However, even though DAP passes quantatively from the small intestine (Mason and White, 1971), cell walls and cell contents of ^{35}S-labeled rumen bacteria are digested to the same extent (Bird, 1972).

Duodenal N in animals fed purified diets containing urea or casein is largely microbial (BCP), whereas it is a mixture of BCP and IUP with purified diets containing plant proteins and with diets consisting entirely of natural feed ingredients. In some experiments apparent absorbability of NAN from the small intestine was similar in animals fed natural diets or purified diets in which urea was the sole source of N (Salter and Smith, 1977; Zinn and Owens, 1982). In studies summarized by Armstrong et al. (1977), apparent disappearance of amino acid nitrogen was 0.70, 0.69, 0.64, and 0.76 for purified diets containing urea, casein, corn gluten meal, and field beans. These data imply that escape amino acid N from corn gluten meal is less digestible, whereas escape amino acid N from field beans is more digestible than microbial amino acid N.

Level of feed intake could affect absorption of N from the small intestine by influencing passage rate of digesta or by adjusting proportionate amounts of microbial and undegraded feed N. Absorption of amino acid N in sheep fed chopped dried grass or pelleted dried grass at 900 or 1400 g/d was 0.06 ± 0.05 greater at the low level of feed intake (Armstrong et al., 1977). On the other hand, Zinn and Owens (1981a,b, 1982) observed that increasing intake of a high-grain diet increased both bypass of dietary protein and apparent absorption of NAN from the small intestine.

Although there are no comparable data with cows, studies with rats and ewes suggest that absorption may be a more efficient process in lactation compared with pregnancy and in pregnancy compared with the nonpregnant, nonlactating state. This is a consequence of an enhanced absorptive area of the gut and an increased partition of cardiac output to the gut (Oldham, 1981, 1984).

The potential for protecting feed protein from degradation in the rumen was discussed in a previous section. In many early experiments, lack of improvements in animal performance was often a consequence of overprotection and decreased intestinal digestibility (Chalupa, 1975a, 1984). In more recent experiments, treatment of a wide range of feedstuffs with formaldehyde substantially increased the amounts of leucine, isoleucine, valine, histidine, arginine, and phenylalanine absorbed from the small intestine (Barry, 1976). Absorption of lysine, threonine, and sulfur-containing amino acids was increased little or in some experiments decreased.

Based upon data summarized in this chapter and by other investigators, the following values are suggested: apparent absorption of NAN, 0.65; true absorption of NAN, 0.75; apparent absorption of amino acids, 0.70; and true absorption of amino acids, 0.80.

Nitrogen Metabolism in the Large Intestine

Postruminal fermentation primarily in the cecum and large intestine of ruminant animals received little attention until the advent of intestinal cannulation. As little as 4 percent of the total organic matter digestion occurs in the cecum plus large intestine with low intakes of forage diets for sheep (Ulyatt et al., 1975a), but with cattle fed at a high level of intake, up to 37 percent of the total energy digestion can occur past the terminal ileum (Zinn and Owens, 1981b). Digestion in the cecum and large intestine can compensate for incomplete digestion in the rumen where residence time, ammonia supply, or pH may limit extent of digestion. This shift in site of fermentation from the rumen to the large intestine can alter energy and amino acid availability for the animal, microbial yield, as well as fecal N loss. Loss of N in feces is typically the greatest source of N loss to a ruminant animal and must be considered in protein metabolism models. Physiology and digestion in the large intestine have been reviewed recently by Ulyatt et al. (1975a), Hoover (1978), Stevens et al. (1980), Wrong et al. (1981), and Ørskov (1982).

Nitrogen enters the cecum plus large intestine from the ileum and by diffusion through the intestinal wall. Input from the ileum consists of undigested feed protein (IUP), indigestible feed protein (IIP), undigested bacterial protein (BCP), plus endogenous N secreted or sloughed from the earlier sections of the intestinal tract (FPN). Amounts of free amino acids or peptides entering the large intestine are insignificant (Clarke et al., 1966). Based on nucleic acid concentrations at various segments of the small intestine, Ben-Ghedalia (1982) suggested that some bacteria may grow in the last half of the small intestine and contribute to the N supply at the end of the ileum. Ileal N (undigested IUP, undigested BCP, and IIP) has been reported to include 45 to 60 percent amino N, 3 to 4 percent nucleic acid N, from 1 to 13 percent ammonia N, and up to 15 percent urea N (Clarke et al., 1966; Coelho da Silva et al., 1972a; Van't Klooster, 1972). The remaining 8 to 40 percent of the total N is presumably hexosamine and mucus glycoprotein.

Urea is present in ileal contents at concentrations from 50 to 100 percent of that in blood. This is derived from diffusion into the small intestine (Hecker, 1971) and suggests that ureolytic bacteria are not prevalent in the small intestine. Urea is rapidly hydrolyzed on entry into the cecum plus large intestine. From 14 to 37 percent of the total urea turnover in sheep has been attributed to urea hydrolysis in the cecum and large intestine (Hecker, 1971; Hogan, 1973; Nolan et al., 1976). Together with degradation of N compounds from undigested feed, bacterial and endogenous sources, hydrolysis of urea that diffuses into the large intestine from the blood stream helps maintain ammonia-N concentrations in the cecum and large intestine between 6 and 27 mM in sheep, although levels below 4 mM have been reported with ruminant animals fed diets containing higher amounts of grain (Williams, 1965; Hecker, 1971; Kern et al., 1974). Sampling methods and sample handling will alter estimates of ammonia concentration of intestinal and fecal matter (Wrong et al., 1981).

Under most feeding conditions more N enters the large intestine from the ileum than leaves as fecal protein (FP) leading to a net absorption of 0.5 to 2 g daily in sheep (Clarke et al., 1966; Hecker, 1971; Ørskov et al., 1971b; Coelho da Silva, 1972a; Thornton et al., 1970) and 0 to 5 g in cattle (Van't Klooster and Boekholt, 1972; Zinn and Owens, 1982). Nevertheless, the amount of nitrogen passing to the terminal ileum per day is highly correlated with the supply excreted in feces (Zinn and Owens, 1982). Nitrogen absorption from the cecum and large intestine into the blood stream or through diffusion to other organs is enhanced by the high large intestinal pH (7 to 9) with roughage rations and is thought to be primarily ammonia. Ammonia can be utilized by bacteria in the large intestine for BCP synthesis, be passively absorbed into the portal blood system, or passed with FP. Diffusion of ammonia is primarily on the nonionized form. Evidence from nonruminants suggests that pH dictates the fate of ammonia, with more ammonia in feces having a lower pH (Down et al., 1972). Increased availability of energy in the large intestine, achieved through infusion of starch, glucose, or sucrose, will increase FP and decrease urinary protein (UP) excretion (Thornton et al., 1970; Ørskov et al., 1971b; Mason et al., 1977). Part of this change is due to an increase in BCP in feces (Mason et al., 1977), and a part of the increase is in the soluble N fraction, probably associated with a decreased fecal pH. Such a shift from N excretion as UP to FP invalidates certain traditional indices of protein value for ruminants, namely apparent digestibility, the concept of biological value, and possibly metabolic fecal N (FPN). Although generally more N enters the large intestine from the ileum than exits as FP, the magnitude of transfer of nitrogen may depend on diet, intake level, animal species, and other factors.

With *in vitro* preparations, active uptake of amino acids by the colon has been demonstrated (Scharrer, 1978). Yet, transport of amino acids to the serosa remains unproven. Several types of reasoning have been used to suggest that amino acids are absorbed from the large intestine. With horses, feeding of urea or infusion of protein into the cecum can increase N retention (Slade et al., 1970; Reitnour and Salsbury, 1972). Disappearance of ^{14}C amino acids (Hoover and Heitman, 1975) or ^{15}N microbial protein (Slade et al., 1971) from the cecum also could reflect amino acid absorption. However, similar results could occur when microbial digestion in the cecum yields ammonia and volatile fatty acids to be absorbed and used by tissues for synthesis of nonessential amino acids. The low concentrations of free amino acids in the cecum and large intestine might be interpreted to suggest that sufficient quantities of amino acids are not available in the free form for absorption. Low concentrations of amino acids in the large intestine reflect the rapid uptake and catabolism of amino acids by intestinal microbes. Wrong et al. (1981) concluded that amino acid absorption from the large intestine, except in the newborn animal, is quantitatively insignificant. Nevertheless, absorbed ammonia becomes available for amination reactions in tissues and urea synthesis for recycling.

From 4 to 37 percent of the total tract DOM digestibility by ruminants occurs in the cecum plus large intestine. High concentrations of volatile fatty acids and branched chain fatty acids reflect fermentation and proteolysis (Hecker, 1971; Kern et al., 1974). With high concentrate rations, lactate production (Kern et al.,

1974) may lower pH or ammonia may become limiting (Williams, 1965). Since infusions of glucose, starch, and gelatin all increase the amount of fecal N as well as the amounts of bacterial components (DAP, RNA) excreted in feces of sheep (Mason et al., 1977), available energy is thought to be the factor limiting BCP synthesis in the large intestine of sheep under most dietary conditions.

Fecal excretion (FP) has been related to (1) intake of nitrogen and (2) either dry matter intake or fecal dry matter output in attempts to estimate (a) true digestibility of fed protein (IP) and (b) the amount of FPN lost by animals to feces. Regression of apparent digestibility of N against N concentration of the diet gives a slope that represents true digestibility of IP. True digestibility values from a number of trials and summaries are listed in Table 15.

True digestibility estimates range from 85 to 95 percent of feed N (Table 15). These are for the total digestive tract, not for specific N component from the small

TABLE 15   Estimates of True Digestibility and Metabolic Fecal Nitrogen

Source	True N Digestibility	FPN (g/kg of DM Intake)	Species	Diet
Schneider, 1947	91	—	Ovine	All
Holter and Reid, 1959	92.9	35		
Holter and Reid, 1959	88.3	31		
Anderson and Lamb, 1967	85.4	21	Ovine	All
Harris et al., 1972	86.6	31	Bovine	Forage
Harris et al., 1972	85.0	21	Bovine	Forage
Harris et al., 1972	90.8	38	Bovine	Forage
Harris et al., 1972	91.8	40	Bovine	Conc.
Stallcup et al., 1975	90.2	36	Bovine	All
Boekholt, 1976	83.3	33	Bovine	Mixed
NRC, 1976	87.7	26	Bovine	Mixed
Swanson, 1977	89.8	29	Bovine	All
Mason and Fredericksen, 1979	92.0	30	Ovine	Forage and mixed
Dror and Tagari, 1980	84.0	29[a] 14[a]	Ovine	All
Preston, 1982	90.3	34	Bovine	Mixed
Waldo and Glenn, 1982	86.1	29	Bovine	Mixed
Calculated from Morrison, 1959				
Green roughages, N = 65	89.0	38	Mixed	Forage
Dry roughages, N = 75	87.0	30	Mixed	Forage
Silages, N = 25	82.8	27	Mixed	Forage
Concentrates, N = 29	95.0	38	Mixed	Conc.
All feeds, N = 197	93.6	35	Mixed	All

[a] Roughage + concentrate.

intestine as in Table 14. Values are surprisingly constant considering the wide variations in digestible energy content and protein sources used in various diets. Fractionation of feces (Mason, 1969) led to the conclusion that true N digestibilities with sheep fed various diets ranged from 73 to 96 percent.

FPN or nondietary fecal N, is the inevitable loss associated with production of feces. For nonruminant animals FPN has been attributed primarily to erosion of the intestinal lining since increased dietary fiber increases FPN (Mukherjee and Kehar, 1949) and feeding of a purified completely digested diet reduces fecal output, and thereby FPN to zero. FPN for nonruminants usually is correlated more closely with fecal output (IOM intake) than DM intake and thereby may be a result of microbial fermentation in the large intestine. With ruminants, part or all of MFN has been regarded as microbial N either synthesized in the large intestine or indigestible BCP passed through from the rumen (Mason and Fredericksen, 1979). When all nutrients are provided to ruminants through infusion of purified, absorbable nutrients, the quantity of feces produced and the amount of FP decline (Ørskov and MacLeod, 1982). Fermentation in the rumen will reduce the amount of potentially digestible material available for fermentation in the large intestine. The amount of BCP synthesized in the rumen or large intestine is normally related to supply of DOM and, at least in the rumen, it should be negatively related to dietary fiber level. Efficiency of microbial growth (BCPFOM), however, is usually higher with diets containing more ADF. Hence, BCP and BCPFOM may change in opposite directions as dietary roughage level is altered. Adding an inert fiber to a diet for calves can alter the relationship of FP to IOM intake (Strozinski and Chandler, 1972), yet FPN appears to be correlated more closely with indigestible organic matter (IOM) output than DM intake (Swanson, 1982). If FPN in ruminants is a combination of (1) microbial residues from the (a) rumen or (b) cecum and large intestine plus (2) indigestible eroded or secreted protein from the digestive tract as suggested in the PDI system of protein evaluation (Waldo and Glenn, 1982), several factors would be needed to estimate its magnitude. These include: (1) site, (2) extent of organic matter digestion, and (3) the amount of indigestible residue pushed through the digestive tract. Chemical subdivision of FPN into microbial versus nonmicrobial fractions does not define the *origin* of the N. Origin is critical in models of protein metabolism. FP, which originates from intestinal tissue, whether or not it is subsequently incorporated into BCP, must be charged against tissue reserves of essential and nonessential amino acids. In contrast, protein synthesized from NPN in the digestive tract is appropriately charged against nonspecific N reserves,

such as plasma urea. No discount for biological value applies to the latter fraction. FPN has been estimated by several procedures. These include (1) the intercept of the plot of apparently digestible protein against dietary protein level, (2) direct measurement with diets having 100 percent true protein digestibility or labelled isotopes as discussed by Strozinski and Chandler (1972), and (3) by enteral infusions of digestible nutrients (Ørskov and MacLeod, 1982). FPN values, being the intercept of the regression of fecal N or estimated by detergent procedures are also listed in Table 15. Estimates range from 21 to 38 g protein per kilogram DM intake. This value has been subdivided into portions for roughage and concentrate portions of the diet by Dror and Tagari (1980) and has been attributed by some workers to the type of diet fed (Institut National de la Recherche Agronomique, 1978). Ørskov and MacLeod (1982) maintained steers and cows with intragastric infusions of digestible nutrients and measured FP and UP losses. Infusions reduced FP but elevated UP loss compared to feeding of N-free diets. This led the authors to conclude that FPN is a result of microbial fermentation somewhere in the digestive tract, and without microbial activity, FPN approaches zero. Yet, UP loss increased in magnitude similar to the decrease in FPN, suggesting that turnover of protein of the intestine results in irreversible loss of N to the system by either one route or another. Differentiation between the two may not be feasible although the combination of FPN and UPN may be more constant. Whether infusions decreased turnover of protein of the digestive tract has not been determined. Indigestible fiber present in the intestine may absorb secretions and abrade more cells from the lumen of the intestine and thereby increase FP.

Based on the more general equations of Harris et al. (1972), digestible protein = (0.84 to 0.92) IPDM − (0.021 to 0.04) (r^2 > 0.90), one can calculate total FP. Combining terms, subtracting from IP, and multiplying by DM intake reveals that fecal protein (FP) = (8 to 16) IP (in kg) + (21 to 40) DM intake (in kg). Here, FPN is calculated as a function of DM intake and ranged from 21 to 40 g protein per kilogram dry matter intake.

Using the assumption that all protein has a true digestibility of 90 percent, Swanson (1977) calculated FPN based on an extensive literature review. He concluded that FPN was 25 to 40 g protein per kilogram dry matter intake or 61.5 g N per kilogram IDM excreted. For calculation by the current system, FPN was assumed to equal 30 g protein per kilogram DM intake. With an assumed dietary DM digestibility of 67 percent, FPN was calculated to be 90 g protein per kilogram IOM (30/0.33). Although this means of estimating FPN is appealing, results may be misleading. Depending on the feedstuff category chosen, FPN estimated by regression can vary by 70 percent (Harris et al., 1972). This could reflect experimental error or could suggest that FPN is not a constant proportion of feed intake or fecal output. Secondly, dry matter intake and protein intake are correlated in most studies, so FPN and true protein digestibility cannot be estimated independently. As illustrated in Table 15, the FPN estimate increases as the estimate of true digestibility of protein increases. Thirdly, true digestibility of protein calculated by regression across feedstuffs generally exceeds values measured with isotopically labeled feed proteins. In conclusion, mathematical separation of fecal nitrogen into that from dietary versus endogenous origin by regression appears variable and without a biological basis. However, some method to subdivide FP into indigestible IP and FPN and make practical accounting of this nitrogen fraction, which totals from 20 to 68 percent of the total nitrogen loss by animals, as described in early studies by Blaxter and Mitchell (1948), is necessary to displace the concept of protein digestibility and generate requirement values in the newer systems of protein metabolism of ruminant animals.

Fecal N consists of 45 to 65 percent amino nitrogen, 5 percent nucleic acid nitrogen, and 3 percent ammonia nitrogen (Coelho da Silva, 1972a; Van't Klooster and Boekholt, 1972; Hogan, 1973). The residual nitrogen consists of partially degraded nucleic acids, bacterial cell walls, and glycoprotein, as well as nitrogen bound to fiber components. Separation by sonication and modified fiber solubility procedures (Mason, 1969) has suggested that 7 to 28 percent of feces is undigested dietary N, 16 to 59 percent is water-soluble N, and 38 to 74 percent is bacterial plus endogenous debris N (Mason, 1969; Mason and Fredericksen, 1979; Plouzek and Trenkle, 1982). The latter fraction can be subdivided, and, according to concentrations of diamino-pimelic acid and ribonucleic acid, is presumably largely bacterial debris, especially bacterial cell walls (Ørskov et al., 1971b; Mason et al., 1977). In conflict with this general concept that bacterial debris comprises a large fraction of the fecal N, some isotope studies with bacterial cell walls indicate that cell walls are readily digested in the ruminant's small intestine (Hogenraad and Hird, 1970; Bird, 1972), and nucleic acid nitrogen concentrations in feces are generally below 5 percent of fecal nitrogen (Coelho da Silva et al., 1972a). These studies would indicate that the amount of intact bacteria in feces is small.

Although subdividing FP is useful to determine true digestibility of protein, compositional analysis of feces does not reveal the point of origin of FP. Liberated, nonutilized N from endogenous secretions of the intestines can be absorbed and excreted in urine or recycled. The amount of N in the BCP fraction of feces could originate from endogenous essential amino acids or from urea cycled to the digestive tract. If energy available to the microbes of the large intestine is the factor that limits microbial protein synthesis, then N in the bacterial plus endogenous debris fraction of feces is not a suitable indicator of endogenous protein loss. Nevertheless, some estimate of the total amount of FN must be calculated to be included in calculations of the total N economy of the ruminant animal. Indigestible BCP synthesized during ruminal fermentation, should logically be charged against nonspecific or N available in the rumen (RAP), but this fraction is not necessarily determined by extrapolation across protein intakes, since with a protein-deficient diet, this fraction may be reduced. It appears more logical to use an intercept estimate of FPN that is not fundamentally based than to underestimate the total N required to replace inevitable losses.

Nitrogen bound to acid-detergent fiber (one index of IIP) comprises from 1 to 75 percent of feed N and has been used to predict apparent N digestibility of heat-damaged forage (Goering et al., 1972; Thomas et al., 1982). Recovery of feed acid-detergent fiber-N in feces, however, differs with feedstuffs and generally ranges from 39 to 90 percent (Goering et al., 1972; Zinn and Owens, 1982). Insolubility in an acid pepsin solution (PIN) has also been employed as an index of indigestibility (IIP) for nonruminants (AOAC, 1980) and ruminants (Goering et al., 1972). Indiscriminate binding of protein or ammonia N to fiber fractions with heating or in the intestinal tract can reduce N availability drastically.

Several implications of the complexity of fermentation in the large intestine are apparent. Protein metabolism schemes must ultimately charge excreted nitrogen against its origin. The quantity of FP to be charged directly against IP due to indigestibility should be to the amount of IP that is truly IIP. Indicators of IIP have been used to predict indigestibility of heat-damaged feeds (Goering et al., 1972; Thomas et al., 1982), but their usefulness for feeds not damaged by heat remains to be determined.

The capacity to recover and recycle nitrogen from the cecum and large intestine gives the ruminant animal a means to alter the efficiency of nitrogen utilization when demands are altered. This means that biological value of N can increase as recycling increases. The magnitude of this adjustment with various feeding conditions must be determined before N utilization in the large intestine and biological value of metabolizable protein can be properly assessed and calculated.

Nitrogen Metabolism in Tissues

At the tissue level, protein nutrition of ruminants involves amino acid metabolism as in nonruminant species. Studies in cattle (Black et al., 1957; Downes, 1961) have shown that the same amino acids are essential in ruminants as in nonruminants since they are not synthesized in tissues in adequate amounts and must be absorbed from the gastrointestinal tract (GIT). A primary difference between ruminant and nonruminant species is that protein quality is dependent upon the availabiity of amino acids leaving the rumen rather than that in the ingested diet. Ruminants undoubtedly require some optimum ratio of amino acids for most efficient utilization of absorbed amino acids, but the understanding of tissue metabolism of amino acids in ruminants has not progressed as much during the past 10 years as has the understanding of protein metabolism within the GIT. Although there is interest and considerable speculation about amino acid requirements of ruminants (Hogan, 1975; Bergen, 1979; Wolfrom et al., 1979), there is limited information on amino acid requirements of ruminant species. The increase in nitrogen balance of sheep (Nimrick et al., 1970), cattle (Fenderson and Bergen, 1972; Richardson and Hatfield, 1978), wool growth (Reis et al., 1973), and lactation in cows (Clark, 1975b) following postruminal administration of certain amino acids suggests that amino acid requirements may be different than the supply from the rumen and that the efficiency of nitrogen utilization in high-producing ruminants can be improved by manipulation of postruminal amino acid supply.

It is recognized that there is a cellular requirement for all the amino acids incorporated into body proteins, but because the nonessential amino acids can be synthesized by certain tissues within the body if sufficient nonspecific nitrogen and carbon precursors are present, only the 10 dietary amino acids essential for the growing rat (EAA) will be considered here. These are leucine (Leu), isoleucine (Ile), valine (Val), sulfur amino acids (S-AA, Met, Cys), phenylalanine and tyrosine (Phe-Tyr), threonine (Thr), tryptophan (Trp), lysine (Lys), arginine (Arg), and histidine (His).

AMINO ACID METABOLISM

A simplified diagram of amino acid metabolism is given in Figure 13.

Free Amino Acid Pools

Most of the amino acids in the body are bound by peptide bonds in proteins. A small portion of the amino acids are free and equilibrate in pools. The major pools of free amino acids are in extracellular and intracellular tissue fluids and blood.

The free EAA in the bloodstream arise from degradation of tissue proteins and absorption from the GIT. Nearly all the absorption occurs in the mucosal cells of the small intestine as free amino acids or as di- and tripeptides. Most of the peptides are hydrolyzed in the intestinal mucosa to free amino acids before passage to the blood. A portion of the amino acids derived from protein digestion in the intestine may be used for protein synthesis or oxidation by the cells of the intestine before they enter the vascular system. The absorbed amino acids are transported by the blood through the portal vein to the liver before being carried to other tissues. Most of the transport is as free amino acids in plasma, but there is evidence of transport of amino acids to tissues as free amino acids in red blood cells and as peptides (McCormick and Webb, 1982). There is some variation in ratio of free amino acids present in plasma and whole blood (Heitmann and Bergman, 1980) reflecting the ability of red blood cells to concentrate certain amino acids. The proportions of amino acids absorbed from the GIT are

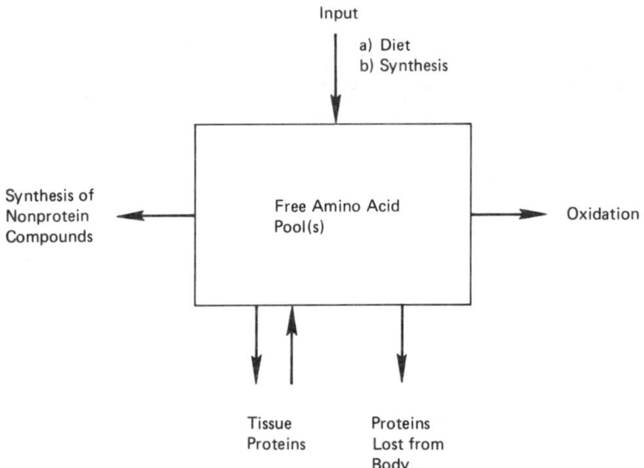

FIGURE 13   Simplified model showing flow of amino acids in mammalian metabolism.

temporarily reflected in the free amino acid pools of plasma after feeding diets that result in large excesses or deficiencies of amino acids passing into the duodenum (Bergen, 1979). Frequently, there is no postprandial rise in plasma amino acids in functional ruminants (Theurer et al., 1966; Fenderson and Bergen, 1972). Between periods of absorption or during fasting, the concentration of EAA increases, that of nonessential amino acid decreases, and the ratios of free EAA more closely reflect the amino acids present in proteins of body tissues.

The concentration of total free amino acids in tissues is 5 to 10 times higher than in plasma, indicating that cells accumulate amino acids against a concentration gradient. Uptake of free amino acids by cells is by active transport across cell membranes, but free amino acids are continuously leaving cells as well (Christensen,

1982). The distribution ratio between free amino acids in tissues and plasma varies widely for various amino acids due to differences in transport systems for different amino acids. When tissues are synthesizing protein, there is a net uptake of amino acids from the blood, but in times of inadequate dietary energy or protein intake there may be a net loss of free amino acids from tissues such as skeletal muscle (Ballard et al., 1976). The extraction of amino acids from the blood by tissues such as the mammary gland may not be in proportion to the appearance of amino acids in proteins (Mepham, 1982). The ratio of amino acids leaving skeletal muscle contains higher proportions of free glutamine and alanine and lower proportions of free branched chain amino acids and glutamic acid than are present in muscle proteins. These differences reflect catabolism of certain amino acids within muscle and the role of alanine and glutamine as a means of transporting ammonia to the liver. The interorgan movement of amino acids and their metabolites may also be beneficial for more adequately meeting the nutritional needs of all body tissues. The concept of "protein reserves" is based upon degradation of protein to amino acids in certain tissues for transport to other tissues for utilization. It has been estimated that the "protein reserves" of the lactating cow can be as high as 27 percent of body protein.

A summary of amino acid extraction by various organs of ruminants is provided in Table 16. Compared with other organs, the mammary gland most efficiently retains the EAA extracted from the blood. The liver removes high proportions of Met, Phe, and Tyr and low proportions of Cys, Val, Leu, and Ile. In sheep there is high umbilical uptake of Val, Leu, Ile, Phe, Lys, and Arg relative to the other EAA (Meier et al., 1981). The extracted Lys and His were retained most efficiently,

TABLE 16   Extraction of Amino Acids by Various Tissues[a]

	Sheep, Liver	Sheep, Kidney	Sheep, Fetus	Calf, Hind Limb	Cow, Mammary Gland
Reference:	Wolff et al. (1972)	Bergman et al. (1974)	Lemons et al. (1976)	McCormick and Webb (1982)	Bickerstaffe and Annison (1974)
Leucine	2.8	2.0	9.8	8.4	42.2
Isoleucine	2.8	0.7	12.2	4.6	38.9
Valine	2.1	2.0	6.8	3.6	26.0
Phenylalanine	20.2	5.6	9.2	2.8	39.8
Tyrosine	16.0	4.4	6.1	4.8	41.8
Threonine	7.7	2.5	6.0	4.4	37.8
Lysine	6.7	5.3	8.9	16.8	58.5
Histidine	8.2	9.0	6.1	7.0	29.4
Cystine	5.8	9.3	—	− 18.4	—
Methionine	15.7	12.7	—	6.4	58.2
Arginine	7.7	11.1	17.7	1.5	53.0

[a]Extraction of free amino acids from plasma. Expressed as: (input–output) ÷ input.

and Val, Leu, Ile, Phe, and Tyr least efficiently by the fetus. Leu and Lys are removed in higher proportion than other amino acids by tissues of the hind limb. These data are based upon plasma free amino acids, which may not be the only source of amino acids to tissues and the degree of extraction would be expected to vary directly with degree of limitation of each amino acid. McCormick and Webb (1982) have reported that amino acids are also extracted from plasma as proteins and peptides and from erythrocytes by the hind limb of calves. Reamination of keto acids also might be a source of amino acids for certain tissues.

Detailed studies have not been conducted with tissues from ruminants, but evidence from rats indicates that labeled amino acids can be incorporated into newly synthesized protein of skeletal muscle and liver without complete mixing with the intracellular amino acid pool, supporting the concept of intracellular compartmentalization of free amino acids. If free amino acids are compartmentalized in cells, withdrawal of amino acids for synthesis and oxidation might not occur from a common pool.

The concept of free amino acid pools is more complex than illustrated in Figure 13. There are many pools of free amino acids in the body that vary in size, ratio of amino acids, and efficiency of amino acid extraction from plasma. Although large quantities of amino acids pass through the free amino acid pools, there is limited storage of free amino acids in the body, and consequently the free amino acid pools do not represent a reserve of amino acids for protein synthesis. Most of the amino acids are bound in proteins and excess amounts of amino acids are oxidized. For efficient utilization of dietary nitrogen, the animal is therefore dependent upon a continuous supply of amino acids of the proper balance.

Utilization of Amino Acids

Removal from the free amino acid pools is mainly for synthesis of body proteins or oxidation. The use of amino acids for gluconeogenesis in the fed ruminant is controversial. Wolff et al. (1972) have suggested that between 11 and 30 percent of the glucose synthesized in fed sheep is derived from amino acids. Bruckental et al. (1980), however, have suggested that amino acids contribute only 1 to 2 percent of the glucose need of the high-yielding cow where glucose and amino acids are both in short supply relative to demand. At any rate, the use of amino acids for gluconeogenesis is only competitive in meeting the protein requirement of the animal if the carbon skeleton of the most limiting EAA is used or if it causes a shortage of precursors for synthesis of nonessential amino acids. Tamminga and Oldham (1980) estimated that no more than one-fourth of the amino acids used for

gluconeogenesis could be from EAA. It is conceivable that feeding excess protein to provide carbon from nonessential amino acids for gluconeogenesis might be beneficial at certain times. Limited amounts of some amino acids are used for synthesis of nonprotein compounds (e.g., creatine, nucleic acids, thyroxine) or excreted in the urine.

Amino acid flux is toward protein synthesis since the Michaelis constants of the enzymes that deaminate amino acids are in the millimolar range, while the enzymes initiating protein synthesis are in the micromolar range. Thereby, when amino acid concentrations are low, greater proportions are bound to synthetic than oxidative enzymes. However, because one or more amino acids or other factors may limit protein synthesis and because free amino acids are transported from one tissue to another by the blood, extraction of amino acids by the liver results in continuous loss of amino acids by oxidation. Since smaller proportions of an amino acid provided in excess are trapped by synthetic enzymes, excess amounts of the amino acid accumulate and plasma concentrations increase. As plasma concentrations increase, the proportion shunted toward oxidation increases.

Protein Synthesis

Synthesis and degradation of body protein is continuous, but proteins in different tissues as well as various proteins within tissues turn over at different rates. In the very young ruminant the largest quantity of protein synthesized is in skeletal muscle (Combe et al., 1979), but increased growth of the GIT associated with consumption of dry feed results in an increased proportion of total protein synthesis in the GIT. Of protein synthesis in sheep (Davis et al., 1981) and cattle (Lobley et al., 1980), 30 to 40 percent of the total synthesis occurs in the GIT, 10 to 20 percent in the skin, 15 to 20 percent in skeletal muscle, and 4 to 8 percent in the liver. The GIT and hide contain about 6 to 20 percent, respectively, of the total body protein but due to rapid turnover account for 30 to 40 percent and 10 to 20 percent, respectively, of total protein synthesized per day. Skeletal muscle, at 40 percent of total body protein, accounts for at least 50 percent of nitrogen retained by a growing animal but only about 20 percent of daily protein synthesis. The fractional rate of protein synthesis is much faster in the GIT, liver, and hide than in skeletal muscle. After the period of rapid growth of the GIT in ruminants, the relative growth rate of the GIT, hide, and liver is less than that of the empty body, but because of high turnover, most of the protein synthesis still occurs in these tissues rather than in skeletal muscle.

As animals mature, the net gain in body protein approaches zero, but large quantities of protein continue to be synthesized due to continued turnover. Lobley et al. (1980) estimated protein synthesis of a mature cow was 1.9 to 3.1 kg per day with 1.0 to 2.1 kg per day occurring in noncarcass components. Large quantities of protein are synthesized in the mammary gland of lactating animals. A cow producing 30 kg of milk containing 3 percent protein secretes 900 g of protein per day. Since there is little degradation of secreted proteins, synthesis probably is only slightly over 900 g per day. It is not known if lactation alters the fractional rate of protein synthesis in other body tissues, but at a minimum, protein synthesis in the noncarcass part of the body must equal that of the mammary gland. The net amino acid requirement for milk protein synthesis, however, is much higher because the proteins are secreted and lost from the body. As tissue proteins turn over, a high proportion of the released amino acids can be reutilized, although efficiency may vary with relationships of proportions of EAA being released and those required for the protein being synthesized. Since hydroxyproline and 3-methylhistidine are not reutilized, their removal reflects turnover rate. Turnover of proteins may account for a greater proportion of the total energy needs than the total amino acid needs of the body.

Synthesis of Nonprotein Compounds

Amino acids are used for the synthesis of a number of nonprotein compounds including creatine, glutathione, carnitine, melanin, dopamine, epinephrine, norepinephrine, thyroid hormones, histamine, carnosine, anserine, taurine, S-adenosylmethionine, nicotinic acid, serotonin, polyamines, γ-aminobutyric acid, purines, pyrimidines, heme, hydroxylysine, and hydroxyproline. The EAA involved include the sulfur amino acids, Arg, Lys, Phe-Tyr, Trp, and His. Only a few of these losses have been quantitated but in total probably account for less than 1 percent of absorbed amino acids. Excretion of creatinine is proportional to body weight and related to the phosphocreatine pool, predominantly in skeletal muscle. Estimates of daily creatinine-nitrogen excretion in cattle and sheep are 3.8 to 9.4 and 8.4 mg per kg body weight per day, respectively (Brody et al., 1934; McLaren et al., 1960). Allantoin-nitrogen, an end product of purine metabolism that is related to digestible organic matter intake and probably reflects absorbed and nonutilized purines from rumen microbes, has been estimated to be 14 mg/kg feed organic matter per day in sheep fed chopped hay but only 0.7 mg/kg feed organic matter per day in sheep given soluble nutrients by intragastric infusion (Antoniewicz and Pisulewski, 1982). In cattle, the excretion of 3-methyl His in the urine is correlated with liveweight and estimated to be 0.6 to 0.7 mg/kg per day (Harris and Milne, 1981). In sheep, the 3-methyl His arising from

degradation of tissue proteins is not quantitatively excreted in the urine (Harris and Milne, 1980).

Excretion of free amino acids from the body in urine is a minor loss under most conditions. There seemed to be no net removal of EAA by the kidney of mature sheep fed at maintenance, fasted, or made acidotic (Bergman et al., 1974).

Amino Acid Oxidation

The major irreversible loss of amino acids from the body is by oxidation. Oxidation of the EAA occurs almost totally in the liver of ruminants. There is considerable catabolism of the branched-chain amino acids in skeletal muscle and other extra-hepatic tissues of nonruminant species, but this does not seem to be the case in ruminants (Coward and Buttery, 1982). Amino acid oxidation in the liver has not been critically studied in ruminants under different nutritional and physiological conditions, but it is known that large portions of free amino acids are removed from blood by the liver (Wolff et al., 1972; Heitmann and Bergman, 1980). In sheep fed at maintenance, nearly all the amino acids added by the portal-drained viscera seemed to be removed from blood plasma by the liver. With greatly reduced absorption of amino acids from the GIT, such as during fasting, removal of amino acids by the liver was maintained. The net escape of amino acids from the liver needs to be reinvestigated in light of erythrocytes and peptides as forms of amino acid transport.

Increasing amino acid intake above requirement increases oxidation. Available evidence suggests that excesses of EAA, due to high absorption from the GIT or by a relative excess due to a scarcity of one or more amino acids, are removed from the free amino acid pools by oxidation in the liver.

Certain proteins represent a direct loss of amino acids from the body. Proteins in hair and scurf, wool, secreted proteins such as milk, proteins secreted or sloughed into the GIT that are not subsequently digested, and proteins retained in the conceptus represent protein losses from the body. Growth of hair and wool requires higher proportions of Val, Leu, Ile, Lys, and Thr and sulfur-containing amino acids as compared with whole body proteins. The amino acids found in higher proportions in milk proteins (Arg, Leu, Ile, and Val) also seem to be more extensively oxidized in the mammary gland as compared with Met, Phe, Tyr, and Trp (Oldham, 1981).

Nitrogen Excretion

Waste nitrogen, principally as urea, arising from deamination of amino acids or ammonia absorbed from the digestive tract, is excreted in the urine, some in milk, or back into the digestive tract. That nitrogen returned to the reticulo-rumen supplements the diet and contributes to the amount of nitrogen available for microbial growth (Cocimano and Leng, 1967; Kennedy and Milligan, 1978; Kennedy et al., 1981, 1982). The amount of urea-N recycled into the rumen appears dependent on the animal and dietary conditions.

Kennedy and Milligan (1980) related clearance of plasma urea to the concentration of rumen ammonia. Their regression developed for cattle fed hay and grain, or hay and sucrose was:

$$Y = 59 - 0.41X + 0.00086X^2;$$

where

Y = clearance of plasma urea in the rumen (ml/h/kg BW), and
X = concentration of rumen ammonia (mg N/L).

To calculate influx, it is then necessary to relate plasma urea concentration to either dietary IP or ruminal ammonia concentrations. More data are available that relate plasma urea concentration to ruminal ammonia concentration. Kennedy and Milligan (1980) found a closer relationship between plasma urea concentration and ruminal ammonia concentration than between plasma urea and dietary crude protein. A linear regression of plasma urea concentration on rumen ammonia concentration was developed from data of Glenn et al. (1983):

$$Y = 79.0 + 14.5X,$$

where

Y = plasma urea-N (mg N/L), and
X = ruminal ammonia-N (mg N/100 ml).

This relationship permits calculation of plasma urea concentration from ruminal ammonia concentration.

Next, ruminal ammonia concentration is needed. This can be estimated from crude protein content and the total digestible nutrient (TDN) content of a diet (Roffler and Satter, 1975a) according to the following equation:

Ruminal NH_3-N (mg N/100 ml)
$$= 38.73 - 3.04\,IP + 0.171\,IP^2 - 0.49\,TDN + 0.0024\,TDN^2;$$
$$R^2 = 0.92,$$

where

IP = dietary crude protein (percent), and
TDN = total digestible nutrients (percent) = 1.02 digestible organic matter (DOM).

From these relationships, the amount of urea-N recycled per kilogram body weight per day could be calculated. Cattle in the study reported by Kennedy and Milligan (1980) were consuming about 2.5 percent of their

body weight daily as dry matter. For this level of intake, amounts of urea-N recycled for diets of various crude protein and DOM contents were calculated. Two regressions were determined:

(1) $$Y = 0.1255 + 0.00426X - 0.003886X^2;$$
$$R^2 = 0.94;$$

where

Y = urea-N recycled (g N/day/kg BW), and
X = dietary IP (percent);

(2) $$Y = 121.7 - 12.01X + 0.3235X^2;$$
$$R^2 = 0.97;$$

where

Y = urea-N recycled (percent of N intake), and
X = dietary IP (percent).

The latter regression is perhaps the most convenient for calculating the amount of urea-N recycled to the rumen. This regression indicates that a diet containing 4 percent IP will lead to urea-N recycled into the rumen equaling 86 percent of dietary N, indicating the significance of this activity in animals fed low-protein diets. For a diet containing 12 percent IP, this value drops to about 25 percent, and for a diet containing 20 percent IP, only 7 percent of the ingested nitrogen is recycled.

Rapidly growing or heavily lactating animals may have lower plasma urea concentrations than the sheep used to develop these urea recycling equations. Tissue or milk synthesis may act as a nitrogen sink, reducing urea synthesis and plasma urea concentration. Highly productive animals might therefore be expected to recycle less urea into the rumen than less productive ruminants fed a comparable diet.

Endogenous protein, from saliva and cells sloughed from rumen epithelium, is an additional source of nitrogen for the rumen microbes, but quantitative information in this area is meager. Furthermore, availability of the nitrogen in keratinized rumen epithelial cells for rumen microbes in unknown. The amount of nitrogen from endogenous protein recycled into the rumen may equal the amount of recycled urea found in highly productive animals.

PROTEIN REQUIREMENTS

The amino acid requirements of ruminants could be estimated by summing the net removal of free amino acids from the free amino acid pools (Figure 12). Practically, this is not possible because all losses have not been quantitated. Because of ease of analysis, the experimental approach used most frequently has been to measure nitrogen rather than amino acid metabolism and convert nitrogen to crude protein (N × 6.25). Nitrogen balance procedures have provided much of the knowledge currently available on protein requirements of animals. Since the body continues to lose nitrogen in the urine and feces, even when dietary intake of nitrogen is nil, these losses were considered to reflect a minimum nitrogen metabolism required to support basic body functions and were termed endogenous (Mitchell, 1962). This parallels energy metabolism with heat production continuing despite starvation. The additional nitrogen metabolism associated with dietary intake of protein has been termed exogenous.

The net protein requirement is the sum of that for maintenance and that expected to be retained in tissues as growth, in the conceptus, wool growth, or excreted in milk. The factorial equation to estimate net protein requirement (g/d) = (FPN + UPN + SPN) + (RPN + YPN + LPN). Requirements for absorbed protein (AP) are determined by assigning metabolic efficiencies for use of absorbed amino acids for various functions.

Requirements for Maintenance

Metabolic Fecal Protein (FPN).   FPN is made up of the undigested fraction of endogenous proteins lost in the feces. Endogenous protein (nitrogen) enters all segments of the GIT. It consists of enzymes, mucus, epithelial cellular debris, serum, lymph, bile, and urea. FPN is considered to represent endogenous proteins lost through the digestive tract as a result of feed intake. Estimates of the quantity of FPN have been made by feeding animals protein-free diets and measuring nitrogen lost in the feces or by feeding diets containing different concentrations of protein and regressing digestible protein against dietary protein to zero protein intake. The latter method usually results in a lower estimate of FPN. In nonruminant species fed low-fiber diets, FPN is related to dry matter intake; however, in ruminants fed diets varying in fiber content it is more closely related to fecal dry matter. In cattle and sheep, FPN ranges from 6 to 8 percent of fecal dry matter. Swanson (1982) has estimated FPN, g/d = 0.068 × fecal dry matter. An alternative estimate, if data on digestibility of the diet are not available, is FPN, g/d = 0.03 × dry matter intake (g/d). Based on this relationship and a DM digestibility of 0.66, FPN = 0.09 × indigestible dry matter (IDM). Mason and Fredericksen (1979) characterized nitrogen fractions in sheep feces and found that much of the fecal nitrogen is microbial debris arising from undigested rumen microbes and from microbial action in the large intestine and cecum. The quantity of nitrogen excreted in the feces increases and that excreted in the urine decreases with increased passage of fermentable substrates to the large intestine (Mason et al., 1981). FPN obviously is of body origin when animals are fed nitrogen-

free diets, but when animals are fed protein, it is not known how much of the nitrogen captured by the microbes in the lower GIT is of body origin and should be considered a true maintenance requirement rather than as a second excretory pathway for waste nitrogen arising from the inefficient use of absorbed nitrogen.

Endogenous Urinary Protein (UPN).   UPN is the nitrogen (protein equivalent) lost in the urine when animals are fed nitrogen-free diets. After feeding nitrogen-free diets for 5 to 7 days, urinary nitrogen is excreted at a relatively constant level, irrespective of the diet fed. Creatine, urea, ammonia, allantoin, uric acid, hippuric acid, and small quantities of amino acids contribute to UPN. UPN is difficult to estimate in ruminants because there is some absorption of amino acids when they are fed nitrogen-free diets as a result of microbial growth originating from nitrogen recycled into the rumen. Swanson (1977) estimated UPN in cattle fed low-protein diets to be UPN, g/d = $2.75 \times wt^{0.5}$. ARC (1980) estimated UPN, g/d in cattle to be: $16.07 \times \ln wt - 42.24$. For sheep, Swanson (1982) estimated UPN, g/d = 1.125 $wt^{0.55}$, and the ARC (1980) estimate for UPN is: 0.1468 $\times wt + 3.375$. More recently Ørskov (1982) has measured loss of nitrogen in the urine of cattle and sheep nourished by intragastric infusion. When nitrogen-free infusates were given, urinary nitrogen losses were 300 to 400 mg $N/wt^{0.75}$, which were considerably higher than nitrogen lost in the urine when ruminants are fed protein-free diets and about triple the estimates above. Animals maintained by intragastric infusion excrete very little nitrogen in the feces, and Ørskov and MacLeod (1982) suggested that metabolic fecal nitrogen measured in feces of ruminants fed nitrogen-free diets is mainly endogenous nitrogen derived from breakdown of tissue protein but incorporated into microbial debris and excreted in the feces.

We are recommending the equations of Swanson (1977, 1982).

Scurf Protein (SPN).   SPN is protein lost from the surface of the body as hair, scurf, and secretions. The estimated loss in cattle is SPN, g/d = $0.2 \times wt^{0.6}$ but is variable depending upon type of hair coat, weather, and ambient temperature.

Requirements for Tissue Growth, Lactation, and Pregnancy

Tissue Protein (RPN).   RPN deposition has been estimated by determination of body composition of growing animals. Many of these studies have been summarized elsewhere (ARC, 1980; Byers, 1982b; NRC, 1984). Net protein gain is a multiple of weight gain and composition of the gain, which are influenced by rate of gain, physiological maturity, previous nutrition, sex, and use of hormonal adjuvants. Three summaries have been made for purposes of estimating net protein requirements of growing cattle by ARC (1980), Robelin and Daenicke (1980), and NRC (1984).

The equation of ARC (1980) to estimate the protein content of empty body gain (EBWG) of cattle of medium frame and gaining 0.6 kg EBWG/d is:

$$\text{kg protein/kg EBWG} = \frac{0.8893\,e^{0.8893\ln\text{EBW}}}{\text{EBW}\,e^{1.1598}}.$$

The correction factors for other types of cattle include a subtraction of 10 percent for small breeds, 10 percent for females, and 1.3 percent for each 0.1 kg/d more than 0.6 kg/d and an addition of 10 percent for large breeds, 10 percent for intact males and 1.3 percent for each 0.1 kg/d gain less than 0.6 kg/d to values calculated for medium steers gaining 0.6 kg/d.

The equations of Robelin and Daenicke (1980) to estimate protein content in EBWG are:

Lipid content of EBW (kg) = L
$$= e^{b_0 + b_1\ln\text{EBW} + b_2(\ln\text{EBW})^2}$$

Daily lipid deposition (kg/d)
$$= 1 = \frac{L}{\text{EBW}}(b_1 + 2b_2\ln\text{EBW})\,\text{EBWG}$$

Daily protein deposition (kg/d) = p
$$= a_0\,a_1\,(\text{EBWG} - 1)\,\text{FFM}^{(a_1 - 1)}$$
FFM = fat free mass = EBW − L

and

	a_0	a_1	b_0	b_1	b_2
Early maturing steers	0.1616	1.060	−6.311	1.8110	0.0000
Early maturing bulls	0.1541	1.060	−1.680	0.0189	0.1609
Late maturing bulls	0.1541	1.060	−5.433	1.5352	0.0000

The equation of NRC (1984) for estimating protein content of shrunk live weight gain (LWG) is:

Daily protein deposition (g/d) = p
= LWG (268 − 29.4 × Mcal energy per kg EBWG).

The discussion and source of those conclusions are in NRC (1984).

For breeds with medium frame and implanted with hormonal adjuvants:

Steers: Retained energy (Mcal/d) = 0.0635 EBW$^{0.75}$
$$\times\ \text{EBWG}^{1.097}$$

Heifers: Retained energy (Mcal/d)
$$= 0.0783\ \text{EBW}^{0.75} \times \text{EBWG}^{1.119}$$

and:

$$\text{EBWG} = 0.956\,(\text{LWG})$$
$$\text{EBW} = 0.891\,(\text{LW})$$

Modifications include:

1. Cattle without hormonal adjuvants contain 5 percent more energy per unit of gain.

2. Medium-frame bulls are equivalent to medium-frame steers of a 15 percent lighter weight.

3. Large-frame animals are equivalent to medium-frame animals of the same sex of a 15 percent lighter weight.

A summary of the application of these three estimates for medium-frame steers of different weights and gaining 0.5, 1.0, and 1.5 kg EBWG per day is given in Table 17. In the 250- to 400-kg weight range, all three methods resulted in similar estimates of net protein requirements. The ARC approach resulted in low estimates for lighter weights and high estimates at the heavier weights. The NRC approach gave high estimates at lighter weights and very low estimates at heavier weights.

It is not certain which equation is most representative of growth of cattle. For medium-frame beef cattle that are fattening, the NRC (1984) method may be most appropriate. Either ARC (1980) or Robelin and Daenicke (1980) is closer to the recommendations for dairy animals approaching maturity without fattening (NRC, 1978). The NRC (1984) equations and modifications have been chosen for use here.

The ARC (1980) equations to estimate the protein content of empty body gain of sheep are:

$$\text{Males: kg protein/kg EBWG} = \frac{0.8995\,e^{0.8995\ln\text{EBW}}}{\text{EBW}\,e^{1.4854}}$$

$$\text{Females: kg protein/kg EBWG} = \frac{0.8164\,e^{0.8164\ln\text{EBW}}}{\text{EBW}\,e^{1.3032}}$$

The protein content of wool is estimated (ARC, 1980) to be protein, g/d = 3 + 0.1 × protein in g/d retained in other issues. A summary of the protein content of gain of sheep is given in Table 18.

Lactating animals often lose weight in early lactation and gain during late lactation and the dry period. Composition (g protein/kg EBW) of weight gain or loss of adult cattle has been estimated to be 175 to 188 (Reid and Robb, 1971) and 160 (NRC, 1978). Protein content of empty body weight changes in adult ewes ranged from 50 to 70 g protein/kg EBW in a study by Rattray et al. (1974).

Products of Conception (YPN). YPN include protein gain in the fetus and growth of the uterus and related tissues. Rattray et al. (1974) and Ferrell et al. (1976) have estimated the protein content of the mammary gland and the gravid uterus during pregnancy of sheep and cattle, respectively. Most protein deposition

TABLE 17 Estimated Net Protein Requirements for Growth of Cattle of Different Body Weights and Gaining at Different Rates

Gain by EBWG/d	Empty Body Weight, kg											
	150 kg			200 kg			250 kg			300 kg		
	NRC[a]	ARC[b]	F[c]	NRC	ARC	F	NRC	ARC	F	NRC	ARC	F
	(g protein per animal per day)											
0.5	101	81	93	92	78	88	83	76	84	74	75	79
1.0	197	151	186	176	147	177	158	143	168	140	140	158
1.5	290	212	279	258	205	265	229	200	252	201	196	237
	350 kg			400 kg			500 kg			600 kg		
	NRC	ARC	F	NRC	ARC	F	NRC	ARC	F	NRC	ARC	F
0.5	66	74	74	59	73	70	44	71	60	29	69	51
1.0	122	138	149	106	136	139	74	133	120	44	130	101
1.5	174	193	223	148	191	209	98	185	180	51	181	151

[a] Estimates derived from NRC (1984).
[b] Estimates derived from ARC (1980).
[c] Estimates derived from Robelin and Daenicke (1980).

in the mammary gland occurs during the last 30 days of pregnancy and is much less than that in the gravid uterus. Estimates of protein deposition in the fetus and uterus (kg/d) of cattle during 141 to 281 days and sheep during days 63 to 147 from conception (ARC, 1980) are:

Cattle: Protein (g/d)
$$= (34.375) \left[e^{(8.5357 - 13.1201e^{-0.00262X} - 0.00262X)} \right]$$

Sheep: Protein (g/d)
$$= (0.0674) \left[e^{(11.3472 - 11.2206e^{-0.00601X} - 0.00601X)} \right]$$

where: X = days post conception.

The daily gain of protein in the products of conception for cattle and sheep are summarized in Tables 19 and 20.

Lactation (LPN). The protein in milk is a multiple of quantity and composition of milk. The LPN requirement (g/d) can be estimated from: Milk N (g/kg) $\times$ 6.25 $\times$ milk yield (kg/d). Total nitrogen of milk includes a nonprotein component that is largely waste products of nitrogen metabolism and when known it may be more correct to use values for true protein content of milk rather than total N $\times$ 6.25. Representative values for true protein content of milk from cattle and sheep are given in Table 21. There is genetic variation in the protein content of cow's milk and the value in Table 21 is more typical of the Friesian breed. There is a relationship between fat and protein content of milk (Overman et al., 1939), and for producers who usually know the fat content of milk, but not true protein, it would possible to estimate protein content from fat content (NRC, 1978).

It is recognized that there is considerable variation in protein content of the products of animal production due to genetics, rate of production, and nutritional his-

TABLE 18 Protein Retention in Gain of Growing Sheep[a]

Empty Body Weight (kg)	Males and Castrates		Females	
	Gain (g/kg Gain)	Wool[a] (g/d/kg Gain)	Gain (g/kg Gain)	Wool[a] (g/d/kg Gain)
10	160	19.0	147	17.7
20	148	17.8	128	15.8
30	142	17.2	119	14.9
40	138	16.8	113	14.3
50	135	16.5	108	13.8

[a] Values for sheep above 10 kg empty body weight and non-Merino breeds.

TABLE 19 Protein Retention in Fetus and Gravid Uterus of Cattle at Different Stages of Gestation

Age (Week from Conception)	Protein Gain (g/d)[a]
20	13.7
22	18.3
24	24.2
26	31.6
28	40.8
30	52.2
32	66.1
34	82.8
36	102.8
38	126.6

[a] Corrected for uterus of nonpregnant cow.

TABLE 20 Protein Retention in Fetus and Gravid Uterus of Sheep at Different Stages of Gestation

Age (Week from Conception)	Protein Gain (g/d)[a]
10	2.4
12	3.9
14	6.3
16	9.5
18	13.9
20	19.5

[a] Corrected for uterus of nonpregnant sheep.

TABLE 21 Protein Content of Milk

Species	g Protein/kg Milk[a]
Cattle	30.0
Sheep	47.9

[a] Corrected for nonprotein nitrogen content of milk (0.55 g N/kg for sheep and 0.30 g N/kg for cattle).

tory, as well as other factors. It is not the intent of this presentation to exhaustively review all of these variables for all classes of ruminants, but rather to present representative data that are needed to estimate protein requirements at the tissue level. Committees for each of the species will need to present more detailed data to more adequately predict protein requirements.

Efficiency of Protein Utilization. The requirement for AP can be determined by correcting the sum of the net protein requirements for maintenance and production by the efficiency with which absorbed amino acids are transferred into product protein. The efficiency with which absorbed amino acids are used for production is difficult to determine, and there are few estimates for producing ruminants. Optimum values for efficiency of amino acid utilization are obtained when protein is limiting production. In addition, there is variation in utilization of different amino acids; the amino acid present in lowest amount relative to requirement is used most efficiently. If one amino acid is limiting, then the utilization of other amino acids will be reduced to some extent related to the deficiency of the limiting amino acid and the relative excess of the other amino acids. Excess amino acids resulting from overfeeding proteins or because of a limiting amino acid are rapidly removed from the body by oxidation and not stored.

Data on efficiency of utilization of mixtures of amino acids that might be representative of absorption are very limited. One approach to estimate these values has been to calculate the biological value of absorbed nitrogen (NRC, 1978, 1984). Estimated efficiencies for growing

cattle range from 0.60 to 0.81 and 0.70 for lactating cows. A similar approach (ARC, 1980) has been to estimate efficiency from: (RPN + UPN)/(IP − FP). With diets limiting in nitrogen, the efficiency for nitrogen use in cattle and sheep is 0.75. It is important to evaluate any efficiency data in the context of the conditions (relative to requirements) that they are gathered.

The two major pathways of amino acid metabolism are protein synthesis or oxidation (Figure 13). Efficiency of transfer of amino acids into product protein can then be calculated from: (Amino acids in product)/(Amino acids in product + Amino acids oxidized) or from: (Amino acid nitrogen in product)/(Amino acid nitrogen in product + Urea nitrogen formed from amino acids in metabolism). This method can be used to determine the efficiency of use of individually labeled amino acids. When the amino acid being studied is limiting production, it is used with high efficiency compared with other amino acids. In calves, utilization of methionine was 0.82 when methionine was limiting growth (Mathers and Miller, 1979). Oldham (1981) and Oldham and Alderman (1982) calculated efficiency of utilization of absorbed amino acids from several studies using urea production to estimate amino acid oxidation and found the values to range from 0.6 to 0.8 for lactating ruminants and from 0.27 to 0.75 for growing ruminants when endogenous urinary nitrogen was included with product nitrogen. Storm et al. (1983) have reported a value of 0.66 for the efficiency of utilization of truly digested bacterial nitrogen for nitrogen retention in lambs.

Based upon the fact that amino acid utilization is lower when protein is fed at or above requirement and amino acid balance usually will be less than maximum, it appears that efficiency of amino acid use should be 0.65 for lactating ruminants and 0.50 for growing ruminants. There is a need for additional research to derive more adequate estimates of efficiency of amino acid utilization, since these values have such a great impact on the calculated requirement for AP.

Additional Roles of Amino Acids. In addition to serving as substrates for protein synthesis, there may be some requirement of amino acids for other needs in the body that under certain conditions might justify feeding additional protein. The role of amino acids in gluconeogenesis has been briefly discussed. Under most practical feeding conditions, it does not appear necessary to feed protein to supply amino acids for synthesis of glucose. The relationships between amino acid metabolism and energy utilization may be economically important with certain ruminant production systems and should be further investigated. Possible roles of amino acids discussed by Oldham (1981) include effects of amino acids on feed consumption, digestion in the rumen, regulation of hormone secretion, and lipoprotein metabolism in the liver.

Application to Ruminant Feeding

INTRODUCTION

Metabolism of nitrogen (N) in the ruminant is defined and reviewed in the several sections preceding this one. No attempt has been made to exhaustively review the literature describing research that has led to the conclusions drawn, although critical and important new contributions are referenced.

The ruminant is unique in its N metabolism in that the active microbial and protozoal populations in the reticulo-rumen modify the composition of the dietary protein (IP) sources *en route* to the absorptive area in the intestine. In addition, the nutrient requirements of the microbial population are not the same as those of the animal. These events result in modified microbial activity and reduced efficiency of the total digestive process (applied to IP). In addition, these processes affect the quantity of amino acids available to the animal and the makeup of the mixture of the amino acids absorbed compared to that in the diet.

Any improvement in the utilization of N by the ruminant ultimately starts with diet formulation, dietary composition in terms of N, energy and other nutrients, and the behavior of the diet in the digestive tract of the animal. This is an important area of research in ruminant nutrition. New principles can be incorporated into the description of the diet, which should encourage further development.

Prior to this publication, protein allowances for ruminants, as reported by NRC, included only amounts of crude protein either to be fed (IP) or digested per 24 h. Although certain guidelines were implied in the use of nonprotein N (NPN), there has been no attempt to deal with other N fractions or with the metabolic dynamics that affect utilization. This report will review current knowledge of N metabolism in the ruminant, present the critical concepts associated with that knowledge, and recommend a method of implementation based on those concepts. This application is designed to be broad and flexible to respond to the ever-increasing understanding of N metabolism by the ruminant and to allow change as needed.

The application of the principles discussed here is organized so that computers can be used to generate solutions. Transfer coefficients and variables have been named so that computer solution can be obtained without using many multiple iterative steps.

NEW CONCEPTS

Several new concepts have been discussed. These can be summarized as they relate to metabolism of N in the ruminant.

Although N may be present in different forms in various pools, all values will be cited in protein (N × 6.25) equivalents to reduce the need for repeated mathematical interconversions.

Dietary protein (IP) can be described in a variety of ways. However, when related to the digestive physiology of the ruminant, three major protein fractions interest nutritionists and producers. Herein these fractions are designated "A," "B," and "C."

The discussion below includes reference to the use of the *in situ* procedure for obtaining estimates of rate and extent of digestion of protein fractions in feeds. Complete discussion of the method can be found in Mehrez and Ørskov (1977) and McDonald (1981). It must also be noted that the *in situ* procedure is only one of several methods for defining the extent and rate of protein degradation in the rumen. Enzymatic procedures and those employing various solvents or detergents may find increased application in the future.

The *in situ* procedure involves the incubation in the rumen of a fistulated animal of a specific amount of feed, in a polyester or nylon bag of pore size (ideally) uniform at 1,500 to 2,000 μ^2. By removal of replicate bags at various times of incubation, the rate and extent of degradation of feed matter can be determined. Mathematical treatment of the data can result in rate constants for digestion and the definition of various chemical fractions of feeds based on their degradation in the rumen.

Concerns often associated with the *in situ* technique include: (1) loss of undegraded proteins that are soluble or become small enough to pass the bag pores with fluid in the rumen or during washing, (2) contamination of residue with attached microbial matter, and (3) the influence of the local environment of the bag on digestion (particle hydration, end product concentration, etc.).

The three protein fractions to be quantitated are:

A. *Rapidly degraded IP*—that fraction of IP that is rapidly converted to ammonia. Included in that fraction is the majority of NPN, free amino acids, and small peptides. The N in this fraction is, for practical purposes, rapidly and almost totally converted to ammonia in the rumen, since the rate of degradation is over 10 times faster than that of passage of solids from the rumen. If ammonia is not incorporated by rumen microbes into protein (BTP) or nucleic acid (NCP), it passes from the rumen (absorbed across the rumen wall or leaves with fluid) and is subject to at least partial loss as urinary urea (UP) or other NPN forms. Whereas many different techniques for measuring this fraction have been suggested, as has been reviewed earlier, the most desirable procedures are either solubility in buffer solutions or incubation *in situ* for 1 to 2 h. Loss of small particles through pores in bags may limit the usefulness of the *in situ* procedure to evaluate this fraction with some feeds. In addition, some slowly degraded but soluble proteins are inappropriately classified in this fraction. Designating fraction A as "soluble protein" frequently causes confusion. Since the absolute quantity is most important, and most diets are mixtures of feedstuffs, it is recommended that when used to describe the diet that this fraction be expressed as a percentage of feed or ration DM, rather than as percentage of IP.

B. *Slowly degraded, available IP*—the difference between total IP and the sum of rapidly degraded (A) plus unavailable IP (C, below). This fraction represents that part of the IP that can potentially escape degradation in the rumen and be available for absorption in the intestine. The extent of degradation of IP in the rumen depends on the residence time of the IP in the rumen. Dietary characteristics and level of feeding both alter the extent of ruminal degradation. Fraction B differs from fraction A in that the rate of degradation of fraction B is of the magnitude of the fractional rate of passage of solids from the rumen. In light of these variables, the expression of the slowly degraded, available IP should be as an absolute quantity, in units of *percentage of ration or feedstuff dry matter*. If rate constants for ruminal degradation are listed, they should be based on measurements made by incubating the feed in question in polyester bags (or other appropriate procedure) for variable lengths of time and fitting regression equation(s) (usually of the general form $Y = A + B^{-dBx}$) to the relationship between X = time and Y = percentage of original slowly degraded, available IP (B) disappearing from the bag (Mehrez and Ørskov, 1977). Fractions A and B must be estimated, although some of B and C will be lost through pores in bags and result in an inflation in the value of "A." The overall calculation of degradation of fraction B should be based on the formula:

$$\text{Degradation} = B * \frac{k_{dB}}{k_{dB} + k_{pB}},$$

where

B = slowly degraded, available IP;
k_{dB} = degradation rate constant; and
k_{pB} = rate of passage from the rumen (measured by the best method available).

It is possible to expand the above equation to incorporate subfractions of "B" and a rate constant appropriate to each. The prediction of degradation of total IP is made according to the equation presented in an earlier chapter. Since most feedstuffs contain a variety of different types of protein, degradation of total protein *in situ* need not necessarily follow first-order kinetics.

C. *Undegraded, unavailable IP*—that fraction that, due either to natural conditions or chemical, heat, or other reactions during processing, is not available to the ruminant by any of the digestive processes and is quantitatively recovered in feces. It behaves as an inert component in any dynamic description of the digestive process. While this fraction is normally associated with silages and forages, many chemical processes can create unavailable IP in nonforage feeds as well. One estimate of unavailable IP is the residue that remains after treatment with acid detergent (Goering and Van Soest, 1972). Questions still to be resolved include the adequacy of acid detergent as a method for quantifying unavailable IP and the impact of this concept on presently accepted protein allowances, as the unavailable IP in feeds is not presently measured. However, it is proposed that until new technology enables a better practical estimate, this is the method of choice. This fraction has a residence time in the rumen similar to feed particles of similar size and specific gravity.

Recycled N (RP)

The role of N recycled into the rumen can be quantitatively important in situations where the microbial requirement exceeds that of the animal as shown by the quantity of N in the diet (i.e., when low-protein diets are fed). While the nonlactating, mature animal is the most common example, at high rates of turnover of rumen contents, more BCP may leave the rumen than would have entered from the diet even at moderately high percentages of IP. This is most apparent when IP is fed in forms that have low "A" fractions and low k_{dB} values or high "C" fractions. As derived earlier, the RP is:

$$Y = 121.7 - 12.01\,X + 0.3235\,X^2; R^2 = 0.97,$$

where

Y = Urea N recycled (percent of N intake), and
X = IPDM (percent of DM).

From IPDM, it is possible to predict how much RP is presented to the rumen. The latter is dependent on saliva flow and composition and concentration of urea N in the blood plasma. Also, the impact of lactation and type of diet (roughage, concentrate) has not been adequately assessed.

The quantity of RP that will be used is based on the factors that govern removal of N from the ammonia pool and is a direct function of the amount of fermented energy that is available in the rumen. The definition and description of the amount of RP is not complete and needs further study. In the development of these recommendations, a constant percentage of IP was considered, recognizing that a single constant would not fit all situations, especially where animals were fed diets very low in protein (IPDM). A value of RP = 0.15 IP fits the lactating dairy cow data reasonably well and is proposed as the factor to use, but it does not fit the data from beef cows fed diets with IPDM of 0.05 to 0.08. In those cases the value for RP would be higher, although precise estimates are not available. The fact that the flow of N from the rumen exceeds intake by an increasing amount at dietary IPDM (percent) of 10 or less suggests that recycling plays an important role.

If one solves the above equation for several IPDM (percent) and calculates RP (percent IP), the following data emerge:

IPDM (percent)	RP (percent IP)	RP (g at 10 kg DM intake)
5	70	350
10	34	340
15	12	180
20	11	220

This illustrates the sensitivity of RP to low IPDM.

The user should be aware that various metabolic pools or "sinks" (lactation, etc.) can alter the RP at a given IPDM (percent), thus making any of the above useful only as estimates. In the beef cow or feedlot steer, solving the equation above for normal IPDM (percent) will suggest diets that undersupply protein needs. Clearly, more work is needed, and on the basis of the significant lack of data, the Committee has chosen the relationship RP = 0.15 IP to allow noniterative and direct solutions to ration formulation, recognizing that in many instances that this value may be in error.

Ruminal Ammonia

Ruminal ammonia-N concentration often serves as an indicator of N-status for microbial production. Roffler and Satter (1975a,b) have presented an equation to predict ruminal ammonia from IP and dietary energy density. This equation was developed for ad-libitum-fed dairy cows fed diets that consisted of commonly fed feedstuffs and may overestimate ammonia N in low-IPDM (percent) diets or other conditions outside the original data set, or those with protein sources more resistant to degradation than soybean meal.

Ammonia concentration represents the residual balance between input and extraction from the ammonia pool in the rumen. Because there is not an equation that contains enough variables to address all of these inputs and balances for all ruminants, ammonia concentration was not part of the calculations used here.

Microbial N Uptake and Efficiency

The quantity of N used in the rumen for microbial synthesis (BCP) is a function of the amount of energy available for microbial growth. While several expressions have been used to relate BCP to fermentable energy in the rumen, the factors that modify the fraction of energy in a ration or feed that is available in the rumen are not well described. Currently, feed analysis reports present an estimate of the energy value of the feed based on the apparent digestibility in the entire digestive tract and when fed at the maintenance level of feeding in many cases (TDN). Until it is possible to predict the fraction of energy actually fermented in the rumen, and the dietary and physiological factors that modify it, it is recommended that BCP be predicted from the following equations, when values preceded by ± are the SE of the coefficient in questions:

Lactating Dairy Cow, Dairy Replacements and All Cattle Fed Diets with 40 Percent or More Roughage:

$$BCP(g) = 6.25\,(-31.86 \pm 10.74 + 26.12$$
$$\pm\,1.30\,TDN); R^2 = 0.77,$$

where

> TDN = consumed TDN (kg), unadjusted for the influence of level of feed intake.

For lactating dairy cows using NEL as the energy unit, an alternative equation is:

$$BCP(g) = 6.25 (-30.93 \pm 10.69 + 11.45 \pm 0.57\,NEL); R^2 = 0.77,$$

where

> NEL = consumed NEL (Mcal), based on intake at three times maintenance as used by NRC (1978).

The relationship between TDN (percent) and NEL (Mcal/kg) is (NRC, 1978):

$$NEL = 0.12 + 0.0245\,TDN.$$

This equation can be used to convert feed analysis results from TDN to NEL as needed or desired.

Cattle Consuming Diets with Less Than 40 Percent Roughage:

$$BCP(g) = 6.25\,TDN (8.63 \pm 1.67 + 14.60 \pm 2.8\,FI - 5.18 \pm 1.37\,FI^2 + 0.59 \pm 0.80\,CI); R^2 = 0.96,$$

where

> TDN = consumed TDN (kg), unadjusted for the influence of level of feed intake;
> FI = forage intake (percent of body weight) (from NRC publications);
> CI = concentrate intake (percent of body weight) (from NRC publications).

Sheep

$$BCP(g) = 6.25 (-1.29 \pm 0.96 + 23.04 \pm 1.71\,TDN); R^2 = 0.73,$$

where

> TDN = consumed TDN (kg), unadjusted for the influence of level of feed intake.

The efficiency with which ruminally available protein (RAP) is trapped by microbes is important in adequately describing the overall metabolism of N in the animal. While the trapping efficiency cannot be 100 percent due to passage of fluid from the rumen that contains RAP and direct absorption of RAP across the rumen wall, there are few data that adequately describe this relationship. It is recognized that as the amount of RAP increases, relative to the energy available in the rumen, the efficiency goes down. However, we cannot define that efficiency at the optimum balance at this time. As a starting point, a *maximum* trapping efficiency of RAP of 0.90 is used here, although BCP synthesis is normally driven by energy availability, not RAP. Future research may allow that constant to be converted to an equation or other variable relationship, especially under conditions of very low IPDM as is found in many rations fed to mature, nonlactating cows.

Intestinal Absorption of N

The various allowances for N by ruminants stated by previous NRC subcommittees have been criticized for presenting apparent N absorption (as digestible protein) data that are not precise due to a variety of modifiers. As a result, the NRC Subcommittee on Dairy Cattle (1978) reported only crude protein. This was done to allow time for refinement of more precise estimates of allowance. The concepts introduced here should better describe the allowances when adequate data become available to validate these concepts. The review of work published previously and presented earlier in this report produces a reasonably consistent value of 0.65 percent as the apparent absorption and 0.75 as true absorption of nonammonia N. The apparent absorption of amino acid N is 0.7 and true absorption is 0.8. It is more useful to partition the components of N into fractions that can be evaluated than to treat N as a single entity, although digestibilities for microbial and undegraded dietary protein (UIP) appear similar. Variable amounts of fraction C will be found in UIP, and thus more variation in digestibility of UIP would be expected.

Fecal N of Nondietary Origin (Metabolic)

The quantity of fecal N that does not result directly from undigested feed or microbial N (FPN) has not been adequately quantitated. Metabolic fecal N represents a major loss of a portion of the dietary N in many feeding instances, particularly the mature ruminant fed near maintenance. It has been common to plot the relationship between N in the diet dry matter (g/kg) and absorbed (apparent) N/diet dry matter (g/kg) to enable an estimate of fecal N at zero IP. Such a plot also produces a slope that has been used to estimate true absorption of N. Reexamination of existing data suggests that there are some deviations from the assumed constancy of the fecal N content from nondietary origin. However, these deviations cannot be expressed as a specific function. If fecal N is plotted against dietary N, both in g/kg DM, diet and physiological status cause marked differences that cannot be related to specific variables at this time. A function based on the quantity of fecal DM necessitates an accurate prediction of that quantity. That can be done if digestion of DM is known. We are recommending that as an *average*, fecal protein of metabolic origin (FPN) be computed from indigestible DM (IDM), which is calculated from TDN. Since TDN percentage declines from the maintenance value (BTDN) as intake increases, and since this decline reflects IDM, we feel that BTDN should be adjusted to an actual value (ATDN) for animals fed diets with more than 40 percent roughage. The NRC (1978) adjusts BTDN downward by 8 percent under the assumption that the dairy cow consumes at three times the maintenance level of intake and the decline in BTDN is 4 percent per multiple of intake equal to maintenance. We recommend this adjustment for computing IDM and FPN for dairy cows.

Thus:

$$ATDN = 0.92\,BTDN,$$

and

$$IDM = (1 - ATDN),$$

where

ATDN and BTDN are fractional values.

It is further assumed that IDM contains 14.4 g N of metabolic origin/kg, or 90 g FPN/kg.

The total requirement of the animal will include the needs for *maintenance protein (SPN + UPN), metabolic fecal protein (FPN), and production (RPN + YPN + LPN)*.

CALCULATION OF DAILY ABSORBED TRUE PROTEIN NEEDED BY ANIMAL

As indicated above, the protein requirement of the animal can be estimated as the sum of three functions: (a) maintenance, (b) obligatory metabolic fecal protein, and (c) production. In a factorial approach, the following relationships can be used to establish the protein needs of the animal, in units of absorbed N $\times$ 6.25 (AP):

A. *MAINTENANCE*:
   Maintenance protein
   = [*scurf protein (SPN) + endogenous urinary protein (UPN)*] $\div$ *0.67)*
   a.1. Scurf protein (g/day) = $0.2\,W^{0.6}$
   a.2. Endogenous urinary protein (g/day)
         = $2.75\,W^{0.5}$ (cattle)
         = $1.125\,W^{0.55}$ (sheep)
   W = body weight (kg)
   0.67 = amount of tissue (maintenance) protein produced from 1.0 g absorbed protein (MPNMPA).

B. *OBLIGATORY METABOLIC FECAL PROTEIN*:
   b.1. *Metabolic fecal protein (FPN) (g/day) = 90 IDM*
         IDM = daily indigestible dry matter excretion (kg), calculated from: DM (1 − ATDN)
         where: ATDN = 0.92 BTDN
                 BTDN = TDN at maintenance, as normally reported from feed analysis laboratories.

C. *PRODUCTION*:
   c.1. *Growth requirement (g/day) = RPN $\div$ 0.50(g/day)*
         0.50 = amount of gained tissue protein produced by 1.0 g absorbed protein (RPNRPA)
         RPN = gain in tissue protein, (g/day), from Tables 16 or 17, or estimated from gain in empty body (digesta free) (EB) by:

Cattle:
RPN (g/day) = LWG (268 − 29.4 Energy/kg EBWG)

where:
LWG = live weight gain (kg)
Energy/kg EBWG = Mcal retained energy (RE)/kg gain in empty body
EBWG = 0.956 LWG
EBW = 0.891 LW (live weight) and:

Steers:
RE(Mcal/day) = $0.0635\,EBW^{0.75}*EBWG^{1.097}$
Heifers: RE(Mcal/day)
   = $0.0783\,EBW^{0.75}*EBWG^{1.119}$ (both of above with medium frame and implanted with hormonal adjuvants)

Modifications to the above:
   (1.) Cattle without hormonal adjuvants contain 5 percent more energy per unit gain;
   (2.) Medium-frame bulls are equivalent to medium-frame steers weighing 15 percent less.
   (3.) Large-frame animals are equivalent to medium-frame animals of the same sex at 15 percent lighter weight.

Sheep:

Males: RPN (g/day)

$$= \left\{\frac{0.8995e^{0.8995\ln EBW}}{EBWe^{1.4854}}\right\}*EBWG \text{ (kg/day)}$$

Females: RPN (g/day)

$$= \left\{\frac{0.8164e^{0.8164\ln EBW}}{EBWe^{1.3032}}\right\}*EBWG \text{ (kg/day)}$$

c.2. *Reproduction requirement (g/day)* = gain in protein in fetus and uterus during second half of gestation (days 141–281, cattle; 63–147, sheep) = [YPN (g/day) ÷ 0.50]
where:
0.5 = amount of uterine and fetal protein produced from 1.0 g absorbed protein (YP-NYPA)
YPN = gain in protein (g/day), as uterine and fetal tissue, from Tables 18 or 19, or estimated from:

Cattle: YPN (g/day)
$$= (34.375) \left[e^{(8.5357 - 13.1201 e^{-0.00262X} - 0.00262X)} \right]$$
X = days from conception between 141 and 281.

Sheep: YPN (g/day)
$$= (0.0674) \left[e^{(11.3472 - 11.2206 e^{-0.00601X} - 0.00601X)} \right]$$
X = days from conception between 63 and 147.

c.3. *Wool growth requirement* (g/day)
= (3.0 + 0.10 RPN) ÷ 0.50
RPN = estimated gain from growth equations for sheep
0.50 = amount of wool protein produced from 1.0 g absorbed protein (SPNSPA)

c.4. *Lactation requirement* (g/day) = LPN ÷ 0.65
0.65 = amount of milk protein produced from 1.0 g absorbed protein (LPNLPA)

D. *PROTEIN LOSS*:
d.1. *Tissue protein mobilization (g/day)* = *160 EBWL*
160 = amount of absorbed protein (g) in 1.0 kg mobilized body tissue
EBWL = empty body weight loss (kg/day).

Total Amount of Absorbed True Protein Needed = (a.1. + a.2. + b.1. + c.1. + c.2. + c.3. + c.4. − d.1.)

CALCULATION OF DAILY NEED OF TRUE PROTEIN IN THE SMALL INTESTINE OF THE ANIMAL

The difference between the amount of absorbed true protein needed by the animal and the amount to be delivered to the small intestine is due to indigestibility and the inefficiency of absorption. As noted in an earlier section, the total disappearance of amino acids from the small intestine and presumed absorption of amino acids is, on the average, 0.80. Thus, in order to provide 0.80 g of absorbed amino acids (protein), 1.00 g of material must be provided to the small intestine:

Protein to Small Intestine (g/day) = Absorbed True Protein Need (g/day) : 0.80.

CALCULATION OF FLOW OF TRUE PROTEIN TO SMALL INTESTINE

The protein flow to the small intestine is the combined sum of microbial protein and the protein in feedstuffs that escapes degradation in the rumen. Certain corrections must be made to equate the protein flow with that needed in the small intestine. First, it is assumed that 80 percent of the microbial crude protein (BCP) is true protein (BTP), and thus 20 percent (nucleic acids, etc.) will not contribute to the absorbed amino acid pool (unless recycled to the rumen, since a large percentage of this N is absorbed). Second, included in the escaped feed protein is the unavailable fraction, C, which passes through the animal undigested. The flow of protein to the small intestine must be corrected for both of these components before they are compared with the amount needed by the animal.

Microbial Protein (BCP) (g/day) =
a. *Lactating Cows and Other Cattle Consuming Diet with More than 40 Percent Roughage* –
BCP = 6.25 (− 31.86 + 26.12 TDN),
or
BCP = 6.25 (− 30.93 + 11.45 NEL).
b. *Cattle Consuming Diets with Less than 40 Percent Roughage* –
BCP = 6.25 TDN (8.63 + 14.60 FI − 5.18 FI² + 0.59 CI).
BTP = 0.80 BCP.
c. *Sheep* –
BCP = 6.25 (− 1.29 + 23.04 TDN)

The variables in the above equations are defined earlier.

Microbial True Protein (BTP) (g/day) = 0.80 BCP.

Feed Protein Escape (g/day)

$$= IP \left(B * \frac{k_{pB}}{k_{dB} + k_{pB}} + C \right)$$

The variables in this equation are defined earlier.

The *quantity* of fraction B (g/day) that escapes is dependent on the rate of passage (k_{pB}) and digestion (k_{dB}) of fraction B. The k_{dB} is variable and depends on the chemical and physical properties of IP and level of feeding and K_{pB}, rate of passage, is variable also. Thus, even though the equation suggests that one can easily compute the IP escape, the variation in the components of the equation makes estimation imprecise.

Some estimates of the amount of protein escaping ruminal degradation can be found. In most cases, the tables of values are more useful for *ranking* of feeds than in actual quantitation, because of the variation noted above and the presence of fraction C. For now, the user is faced with the need to choose a value for IP escape based on limited current data.

CALCULATION OF AMOUNT OF NITROGEN AVAILABLE IN THE RUMEN FOR MICROBIAL SYNTHESIS

The amount of N available for BCP in the rumen is the sum of the N from DIP and that recycled into the rumen as urea or other soluble sources in saliva (RP). Whether this N is incorporated into BCP is a function of energy supply, as noted above. A further set of consequences of the microbial growth process are: (a) that only 80 percent of the N trapped in BCP is amino acid N (BTP) (thus, the overall process is no more than 80 percent efficient) and (b) that the efficiency of trapping N (ammonia) from rumen fluid is less than 100 percent (assumed to be 90 percent here), due to flux of ammonia with fluids to the omasum. Efficiency probably approaches 100 percent at very low concentrations of ammonia and drops below 90 at higher concentrations. Hence, no more than 72 percent of the nitrogen from a protein degraded in the rumen can be expected to be recovered as BTP. Hence, RP (primarily as urea) becomes important in the nitrogen economy of the animal.

Recycled nitrogen (RP) (percent of intake) can be predicted from dietary crude protein percentage by:

RP = 121.7 − 12.01 IPDM + 0.3235 $IPDM^2$; R^2 = 0.97. This is an iterative process. The alternative is to use 0.15 IP in a direct solution, which we recommend.

Degraded feed protein (DIP) (g/day)

$$= IP\left(A + B * \frac{k_{dB}}{k_{dB} + k_{pB}}\right)$$

IP, A, B, k_{dB}, k_{pB} are defined above.

An alternative would be to estimate the quantity of degraded protein from values in tables comparing feeds. Estimates of degradation are subject to the errors of escape protein, discussed above.

Thus:
*Protein (available in rumen (RAP) g/day) = (RP * IP) + DIP*

When comparing protein available in the rumen with microbial protein:
Maximum microbial protein (BCP) < 0.9 RAP

The user should be aware that the conversion of available N in the rumen to microbial protein is here assumed to have a maximum efficiency of 0.90.

The above represents a set of approximations, meaning that once the need is calculated, and a sample diet is balanced, it must be checked and modified to ensure that the inputs meet needs of the animal.

The material on the following page represents an example of a form that can be used to set up and complete the calculation of the protein needs of an animal and the dietary characteristics which best meet those needs, based on the information presented herein. Further examples and tables can be found in the Appendix tables.

EXAMPLE AND FORM FOR CALCULATING PROTEIN NEED AND DIETARY PROTEIN CHARACTERISTICS

A. *Example*: 600-kg BW dairy cow, 30 kg 3.5 percent fat milk, 3.3 percent protein, 150 days pregnant, + 0.10 kg/day body weight change.

B. *Requirements*:
1. *Maintenance* = [SPN + UPN] ÷ 0.67
   a. SPN = $0.2BW^{0.6}$ =   (9.3 g)
   b. UPN = $2.75BW^{0.5}$ =   (67.4 g)
   c. [SPN + UPN] ÷ 0.67 =   *115* g
2. *Metabolic Fecal Protein* = FPN = 90 IDM
   a. BTDN = BTDNM + BTDNL
      BTDNM = $0.0352\ BW^{0.75}$ =   (4.27 kg)
      BTDNL = (Milk, kg) (NRC TDN/kg milk)
            = (30) (0.302) =   (9.06 kg)
      BTDN = (4.27) + (9.06) =   (13.33 kg)
   b. ATDN = 0.92 BTDN =   (12.26 kg)
   c. DM = BTDN/NRC BTDNDM
            = 13.33/0.75 =   (17.77 kg)
   d. ATDN ÷ DM   (0.69)
   e. IDM = DM (1 − ATDN ÷ DM) = (5.51 kg)
   f. FPN = 90 (5.51) =   496 g
3. *Production* = (RPN ÷ 0.50) + (YPN ÷ 0.50) + (LPN ÷ 0.65)
   RPN:
   a. Use large frame, no hormonal adjuvants
   b. Adjustment for frame
           = 600 × (1 − 0.15) =   (510 kg)
   c. EBW = 0.891 (510) =   (454 kg)
   d. EBWG = 0.956 (0.10) =   (0.096 kg/day)
   e. RE (Mcal/day) = $0.0783(454)^{0.75}$
      $(0.096)^{1.119}$ =   (0.56 Mcal/day)
   f. RE adjustment for no hormones
           = RE 1.05 =   (0.59 Mcal/day)
   g. RE (Mcal/kg EBWG)
           = 0.59 ÷ 0.096 =   (6.12 Mcal)

h. RPN (g/day)
      = 0.10 (268 − 29.4 (6.12)) =   (8.8 g)
i. RPN & 0.50 =   17.6 g
YPN:
a. YPN (g/day)
      = (34.375)
         $[e^{(8.5357 − 13.1201e^{0.000262(150)} − 0.00262(150))}]$
      = (34.375) (0.4895) =   (16.8 g)
      (*Note*: extrapolation from Table 18 = 16.9 g
b. YPN ÷ 0.50 =   33.6 g
LPN:
a. Milk protein = (30) (0.033) (1000) = (990 g)
b. LPN ÷ 0.65 =   1523 g
Total Requirement for Absorbed Protein (AP):
AP = (115) + (496) + (17.6) + (33.6)
      + (1523) =   2185.2 g
C. Production of Bacterial Protein (BCP):
(Assume that diet more than 40 percent roughage)
BCP (g) = 6.25 (− 31.86 + 26.12 (13.33)) = 1977 g
D. Bacterial True Protein (BTP):
BTP(g) = 0.80 BCP =
      = 0.80 (1977) =   1581 g
E. Ruminally Available Protein (RAP):
RAP (g) ≥ BCP ÷ 0.90
      ≥ (1977) ÷ 0.90 ≥   2196 g
F. Digested Bacterial True Protein (DBP):
DBP = 0.80 BTP =
      = 0.80 (1581) =   1265 g
G. Digestible Undegraded Intake Protein (DUP):
DUP = AP − DBP
      = (2185.2) − (1265) =   920.2 g
H. Undegraded Intake Protein (UIP):
UIP = DUP ÷ 0.80 =
      = (920.2) ÷ 0.80 =   1150 g
I. Small Intestine True Protein Flow (STP):
STP = BTP + UIP
      = (1581) + (1150) =   2731 g
J. Intake Protein (IP):
(Use 15 percent of IP as RP)
IP = (RAP + UIP) ÷ 1.15   IP = DIP + UIP
   = ((2196) + (1150)) ÷ 1.15 =   2910 g
K. Intake Protein in Diet Dry Matter (IPDM):
IPDM = (IP) ÷ (1000 DM)
      = (2910) ÷ (17770) =   0.163 g
                                      = 16.38 percent
L. Undegraded Protein Needed in Diet (UIPIP):
UIPIP = UIP × IP
      = (1150) ÷ (2910) =   0.395
                                      = 39.5 percent
M. Degraded Protein Needed in Diet (DIPIP):
DIPIP = DIP ÷ IP
      = (2913 − 1150) ÷ (2910) =   0.605
                                      = 60.5 percent
DIP = IP − UIP

DIP = (RAP + UIP) ÷ 1.15 − UIP

Utilization of Nonprotein Nitrogen (NPN)

Originally, interest in defining many of the parameters associated with ruminant nitrogen usage dealt with ways to predict the usefulness of NPN. Many publications have been written on that subject.

This subcommittee feels that the system that has been presented, complex as it may seem to be, represents a quantitative evaluation of the entire set of conditions under which NPN can be used, and how much. By defining the quantity of the dietary protein that must be degraded in the rumen to meet the need for microbial growth, the potential for reduced intake and digestion should be avoided. On the other hand, by defining the total amount of protein that must leave the rumen to meet the animals' needs, the user is in a position to predict when NPN can be used to help achieve those needs.

Based on the equation and relationships developed in this publication, a set of tables (Appendix Tables 4 to 6) are presented as *guidelines* for determining those dietary and production conditions under which additional NPN would not be expected to be utilized by the rumen microbial population. In addition, Appendix Tables 7 and 8 present data, computed from these same concepts, on the concentration of dietary protein needed for a variety of conditions for beef cattle as well as the percentage of that protein that should escape ruminal degradation to result in the optimum feeding program for that animal. These latter tables can also be used to evaluate the potential for using NPN and to aid in selecting sup plemental protein sources.

Computer Programs

It is recognized that many users of this publication will not be in a position to use a computer program at this time. The number of opportunities for computer application will certainly increase in the future, however. In addition, many advisors, extension specialists, and industry personnel, plus those in teaching and research, use computers routinely and increasingly in the formulation and evaluation of rations and feeding programs.

In order to anticipate the increased dependence on the computer, and to present a rigorous model to evaluate the concept presented here, the subcommittee has chosen to provide Fortran IV programs for the calculation of the dairy (Appendix 9) and beef (Appendix 10) applications. These programs are presented with appropriate commentary and explanation to allow one to use them with little difficulty. In addition, there is an increasing number of published microcomputer programs, spreadsheet applications, etc. (Lane and Cross,

1985) that will enable the user to apply these concepts easily. It is anticipated the microcomputer application will be the common mechanism of use, and the reader is thereby encouraged to pursue that avenue.

Unresolved Problems and Some Areas Needing More Research

During the course of the deliberations of this subcommittee, many areas of ruminant N metabolism were found to be poorly defined or were defined in specific narrow conditions that did not allow application to all classes. We feel that these are some of the areas that need research attention.

A. *Recycled N.*   The data here are both meagre and questionable in their application to normal or practical diets. Whereas we recognize that at low IPDM, RP is of great importance, application of the relationship presented in which RP is a function of IPDM results in unreasonably low IPDM for animals at low production levels. As a result, we present the ratio approach (RP = 0.15 IPDM) as an estimate.

B. *Efficiency of Microbial Uptake of RAN.*   We know that this cannot be 100 percent as long as RAN can leave the rumen on a continual basis with fluids, etc. We also know that when RAN is in excess of that which can be converted to BCP, the efficiency is low. However, when RAN is supplied in amounts intended to minimize waste and maximize BCP yield at prevailing dietary non-N circumstances, the biological efficiency is not clear. We have chosen 90 percent as an estimate and hope that more quantitative data will emerge from future research.

C. *Prediction of Microbial Yield (BCP).*   There are many data on this subject, gathered by a variety of techniques. In the process of developing a set of predictors that can be driven from dietary information that is available for practical use, the picture is less clear. While we have resorted to a whole-gut measure of energy, knowing that this is subject to many animal and dietary factors, the alternatives are not clear. A review of the variation around some of the coefficients in the prediction equations for BCP will point out the lack of precision. In order to construct a system that is driven from commonly measured (or predicted) energy measurements at the level of the rumen, much work is needed on the appropriate relationships.

D. *Transfer Coefficients.*   In addition to the BCPRAP relationship noted above, there is a need for more data on the other N transfers that take place in ruminants. While some term describing "Biological Value" is desirable in defining the N metabolism of all organisms, it is not possible to make such a jump with ruminants. For example, the assumed values for LPNLPA (0.65), RPNRPA (0.50), and MPNMPA (0.65) are based on some data and "best estimates." It is recognized that the balance in available amino acids (AP) is going to have an impact on the transfer coefficients and that the sensitivity of these impacts will depend on the number of metabolic options available to the animal. As more emphasis is given to IP that escapes rumen degradation (UIP), the amino acid balance of the UIP becomes important in evaluating the transfer coefficients. Formulation of diets only on the basis of how much AP is presented to the animal will in some cases be inappropriate because of poor distribution of essential amino acids in the AP. Particular attention needs to be given to lysine and methionine. Until specific data are available, further refinement is not possible.

This subcommittee presents the above document as an attempt to improve the mechanisms for rationing of nitrogen and protein for ruminants, based on the current knowledge. It is hoped that subsequent revisions will be able to build on and advance these concepts.

References

Agricultural Research Council. 1980. The Nutrient Requirements of Ruminant Livestock. Slough, England: Commonwealth Agricultural Bureaux.

Aguirre, E. O., A. L. Goetsch, and F. N. Owens. 1984. Fermented corn grain—site and extent of nutrient digestion in steers. Okla. Agric. Exp. Res. Rep. MP-116.

Akin, D. E., G. L. R. Gordon, and J. Hogan. 1983. Rumen bacterial and fungal degradation of *Digitaria pentzii* grown with and without sulfur. Appl. Environ. Microbiol. 46:738.

Allison, M. J. 1970. Nitrogen metabolism of ruminal micro-organisms. Pp. 456–473 *in* Physiology of Digestion and Metabolism in the Ruminant, A. T. Phillipson, ed. Newcastle-upon-Tyne: Oriel Press.

Alwash, A. H., and P. C. Thomas. 1971. The effect of the physical form of the diet and the level of feeding on the digestion of dried grass by sheep. J. Sci. Food Agric. 22:611.

Ammerman, C. B. 1973. Effect of processing on the nutritional value of dried citrus pulp. P. 297 *in* Effect of Processing on the Nutritional Value of Feeds. Washington, D.C.: National Academy of Sciences.

Ammerman, C. B., and S. M. Miller. 1974. Selenium in ruminant nutrition. A review. J. Dairy Sci. 58:1561.

Anderson, M. J., and R. C. Lamb. 1967. Predicting digestible protein from crude protein. J. Anim. Sci. 26:912 (Abstr.).

Annison, E. F., D. Lewis, and D. B. Lindsay. 1959. The metabolic changes which occur in sheep transferred to lush spring grass. I. Changes in blood and rumen constituents. J. Agric. Sci., Camb. 53:34.

Anrique, R. 1976. Body composition and efficiency of cattle as related to body type, size, and sex. Ph.D. dissertation. Cornell University.

Antoniewicz, A. M., and P. M. Pisulewski. 1982. Measurement of endogenous allantoin excretion in sheep urine. J. Agric. Sci., Camb. 98:221.

Arambel, M. J., and C. N. Coon. 1981. Effect of dietary protein on amino acids and microbial protein of duodenal digesta. J. Dairy Sci. 64:2201.

Armstrong, D. G., and K. Hutton. 1975. Fate of nitrogenous compounds entering the small intestine. P. 432 *in* Digestion and Metabolism in the Ruminant, I. W. McDonald and A. C. I. Warner, eds. Armidale, NSW, Australia: The University of New England Publishing Unit.

Armstrong, D. G., G. P. Savage, and D. G. Harrison. 1977. Digestion of nitrogenous substances entering the small intestine with particular reference to amino acids in ruminant livestock. P. 55 *in* Proc. 2nd Int. Symp. on Protein Metabolism and Nutrition. EAAP Pub. 22.

Wageningen: Centre for Agricultural Publishing and Documentation.

Association of Official Analytical Chemists (AOAC). 1980. Official Methods of Analysis. 13th ed. Washington, D.C.: Association of Official Analytical Chemists.

Balch, C. C., and R. C. Campling. 1965. Rate of passage of digesta through the ruminant digestive tract. P. 108 *in* Physiology of Digestion in the Ruminant, R. W. Dougherty, ed. Washington, D.C.: Butterworths.

Baldwin, R. L., and S. C. Denham. 1979. Quantitative and dynamic aspects of nitrogen metabolism in the rumen: A modeling analysis. J. Anim. Sci. 49:1631.

Baldwin, R. L., L. J. Koong, and M. J. Ulyatt. 1977a. A dynamic model of ruminant digestion for evaluation of factors affecting nutritive value. Agric. Systems 2:255.

Baldwin, R. L., L. J. Koong, and M. J. Ulyatt. 1977b. The formation and utilization of end-products: mathematical models. P. 347 *in* Microbial Ecology of the Gut, R. T. J. Clarke and T. Bauchop, eds. New York: Academic Press.

Ballard, F. J., O. H. Filsell, and I. G. Jarrett. 1976. Amino acid uptake and output by the sheep hind limb. Metabolism 25:415.

Barnard, E. A. 1969. Biological functions of pancreatic ribonuclease. Nature 221:340.

Barry, T. N. 1976. The effectiveness of formaldehyde treatment in protecting dietary protein from rumen microbial degradation. Proc. Nutr. Soc. 35:221.

Bartley, E. E., E. L. Herod, R. N. Bechtle, D. A. Saplenza, and B. E. Brent. 1979. Effect of monensin or lasalocid with and without niacin or amicloral on rumen fermentation and feed efficiency. J. Anim. Sci. 49:1066.

Bauchop, T. 1981. The anaerobic fungi in rumen fibre digestion. Agric. Environ. 6:339.

Bauchop, T., and S. R. Elsden. 1960. The growth of microorganisms in relation to their energy supply. J. Gen. Microbiol. 23:457.

Beever, D. E., and D. J. Thomson. 1981. The effect of drying and processing red clover on the digestion of the energy and nitrogen moieties in the alimentary tract of sheep. Grass Forage Sci. 36:211.

Beever, D. E., D. J. Thomson, and D. G. Harrison. 1971. The effects of drying and the comminution of red clover on its subsequent digestion by sheep. Proc. Nutr. Soc. 30:86A.

Beever, D. E., D. J. Thomson, and S. B. Cammell. 1976. The digestion of frozen and dried grass. J. Agric. Sci., Camb. 86:443.

Beever, D. E., J. L. Black, and G. J. Faichney. 1980–1981. Simulation of the effects of rumen function on the flow of nutrients from the stomach of sheep. 2. Assessment of computer predictions. Agric. Systems 6:221.

Beever, D. E., D. F. Osbourn, S. B. Cammell, and R. A. Terry. 1981. The effect of grinding and pelleting on the digestion of Italian ryegrass and timothy by sheep. Br. J. Nutr. 46:357.

Ben-Ghedalia, D. 1982. Fate of digesta ribonucleic acid in small intestine of sheep. J. Dairy Sci. 64:2422.

Ben-Ghedalia, D., H. Tagari, A. Bondi, and A. Tadmor. 1974. Protein digestion in the intestine of sheep. Br. J. Nutr. 31:125.

Bergen, W. G. 1978. Postruminal digestion and absorption of nitrogenous components. Fed. Proc. 37:1223.

Bergen, W. G. 1979. Free amino acids in blood of ruminants—philosophical and nutritional regulation. J. Anim. Sci. 49:1577.

Bergen, W. G., E. H. Cash, and H. E. Henderson. 1974. Changes in nitrogenous compounds of the whole corn plant during ensiling and subsequent effects on dry matter intake by sheep. J. Anim. Sci. 39:629.

Bergen, W. G., J. R. Black, and D. G. Fox. 1979. Upper limit of ruminal microbial protein synthesis determined for NPN utilization in net protein systems. Feedstuffs, February 19, 1979.

Bergen, W. G., D. B. Bates, D. E. Johnson, J. C. Waller, and J. R. Black. 1982. Ruminal microbial protein synthesis and efficiency. P. 99 in Protein Requirements for Cattle: Proceedings of an International Symposium, F. N. Owens, ed. MP-109. Stillwater, Okla.: Division of Agriculture, Oklahoma State University.

Bergman, E. N., and R. N. Heitman. 1978. Metabolism of amino acids by the gut, liver, kidneys and peripheral tissues. Fed. Proc. 37:1228.

Bergman, E. N., C. F. Kaufman, J. E. Wolff, and H. H. Williams. 1974. Renal metabolism of amino acids and ammonia in fed and fasted pregnant sheep. Am. J. Physiol. 226:833.

Bickerstaffe, R., and E. F. Annison. 1974. The metabolism of glucose, acetate, lipids and amino acids in lactating dairy cows. J. Agric. Sci., Camb. 82:71.

Bird, P. R. 1972. Sulphur metabolism and excretion studies in ruminants. VI. The digestibility and utilization by sheep of ^{35}S from ^{35}S labelled ruminal microorganisms. Aust. J. Biol. Sci. 25:195.

Black, A. L., M. Kleiber, A. H. Smith, and D. N. Stewart. 1957. Acetate as a precursor of amino acids of casein in the intact dairy cow. Biochem. Biophys. Acta 23:54.

Black, J. L., D. E. Beever, G. J. Faichney, B. R. Howarth, and N. McC. Graham. 1980-1981. Simulation of the effects of rumen function on the flow of nutrients from the stomach of sheep. 1. Description of a computer program. Agric. Systems 6:195.

Blackburn, T. H. 1968. Protease production by *Bacteroides amylophilus* strain H18. J. Gen. Microbiol. 53:27.

Blackburn, T. H., and P. N. Hobson. 1962. Further studies on the isolation of proteolytic bacteria from the sheep rumen. J. Gen. Microbiol. 29:69.

Blackburn, T. H., and W. A. Hullah. 1974. The cell-bound protease of *Bacteroides amylophilus* H18. Can. J. Microbiol. 20:435.

Blaxter, K. L., and H. H. Mitchell. 1948. The factorialization of the protein requirements of ruminants and of the protein values of feeds, with particular reference to the significance of the metabolic fecal nitrogen. J. Anim. Sci. 7:351.

Boekholt, H. A. 1976. Nitrogen metabolism of the lactating cow and the role of gluconeogenesis from amino acids. Meded. Landbouwhogesch. Wageningen 76:10.

Brandt, V. M., K. Rohr, and P. Lebzien. 1981. Bietrage zur quantifizierung der N-umsetzungen in den vormgen von milchkuhen. Z. Tierphysiol., Tierernaehr. Futtermittelkd. 46:49.

Braund, D. G., K. L. Dolge, R. I. Goings, and R. L. Steele. 1978. Method of formulating dairy cattle rations. U.S. Patent 4,118,513.

Bray, H. G., and K. White. 1966. Kinetics and Thermodynamics in Biochemistry. New York: Academic Press.

Broderick, G. A. 1978. *In vitro* procedures for estimating rates of ruminal protein degradation and proportions of protein escaping the rumen undegraded. J. Nutr. 108:181.

Broderick, G. A. 1982. Estimation of protein degradation using *in situ* and *in vitro* methods. P. 72 in Protein Requirements for Cattle: Proceedings of an International Symposium, F. N. Owens, ed. MP-109. Stillwater, Okla.: Division of Agriculture, Oklahoma State University.

Broderick, G. A., and W. M. Craig. 1980. Effect of heat treatment on ruminal degradation and escape, and intestinal digestibility of cottonseed meal protein. J. Nutr. 110:2381.

Brody, S., R. C. Procter, and U. S. Ashworth. 1934. Growth and development—with special reference to domestic animals. XXXIV. Basal metabolism, endogenous nitrogen, creatinine and neutral sulfur excretions as functions of body weight. Mo. Agric. Exp. Stn. Res. Bull. 220.

Bryant, M. P., and I. M. Robinson. 1963. Apparent incorporation of ammonia and amino acid carbon during growth of selected species of ruminal bacteria. J. Dairy Sci. 46:150.

Bruckental, I., J. D. Oldham, and J. D. Sutton. 1980. Glucose and urea kinetics in early lactation. Br. J. Nutr. 44:33.

Burroughs, W., N. A. Frank, P. Gerlaugh, and R. M. Bethke. 1950. Preliminary observations upon factors influencing cellulose digestion by rumen micro-organisms. J. Nutr. 40:9.

Burroughs, W., A. H. Trenkle, and R. L. Vetter. 1971. Some new concepts of protein nutrition of feedlot cattle. Vet. Med. Small Anim. Clin. 66:238.

Burroughs, W., A. Trenkle, and R. L. Vetter. 1974. A system of protein evaluation for cattle and sheep involving metabolizable protein (amino acids) and urea fermentation potential of feedstuffs. Vet. Med. Small Anim. Clin. 69:713.

Burroughs, W., D. K. Nelson, and D. R. Mertens. 1975a. Evaluation of protein nutrition by metabolizable protein and urea fermentation potential. J. Dairy Sci. 58:611.

Burroughs, W., D. K. Nelson, and D. R. Mertens. 1975b. Protein physiology and its application in the lactating cow: The metabolizable protein feeding standard. J. Anim. Sci. 41:933.

Byers, F. M. 1982a. Nutritional factors affecting growth of muscle and adipose tissue in ruminants. Fed. Proc. 41:2562.

Byers, F. M. 1982b. Protein growth and turnover in cattle: Systems for measurement and biological limits. P. 141 in Protein Requirements for Cattle: Proceedings of an International Symposium, F. N. Owens, ed. Stillwater, Okla.: Division of Agriculture, Oklahoma State University.

Calderon Cortes, J. F., J. J. Robinson, I. McHattie, and C. Fraser. 1977. The sensitivity of ewe milk yield to changes in dietary crude protein concentration. Anim. Prod. 24:135.

Castle, E. J. 1956. The rate of passage of foodstuffs through the alimentary tract of the goat. I. Studies on adult animals fed on hay and concentrates. Br. J. Nutr. 10:15.

Chalmers, M. I., D. P. Cuthbertson, and R. L. M. Synge. 1954. Ruminal ammonia formation in relation to the protein requirement of sheep. I. Duodenal administration and heat processing as factors influencing fate of casein supplements. J. Agric. Sci., Camb. 44:254.

Chalupa, W. 1972. Metabolic aspects of non-protein nitrogen utilization in ruminant animals. Fed. Proc. 31:1152.

Chalupa, W. 1975a. Rumen bypass and protection of proteins and amino acids. J. Dairy Sci. 58:1198.

Chalupa, W. 1975b. Amino-acid nutrition of growing cattle. P. 175 in Tracer Studies on Non-protein Nitrogen for Ruminants II. Proc. Res. Coord. Mtg. and Panel. Vienna: International Atomic Energy Agency.

Chalupa, W. 1976. Degradation of amino acids by the mixed rumen microbial population. J. Anim. Sci. 43:828.

Chalupa, W. 1978. Digestion and absorption of nitrogenous compounds in ruminants. P. 211 *in* Proc. 3rd World Congr. on Animal Feeding. Madrid.

Chalupa, W. 1980a. Methods for estimating protein requirements and feed protein values for ruminants. Feedstuffs 52:(26)18.

Chalupa, W. 1980b. Chemical control of rumen microbial metabolism. P. 325 *in* Digestive Physiology and Metabolism in Ruminants. Proc. 5th Int. Symp. on Ruminant Physiology, Y. Ruckebusch and P. Thivend, eds. Lancaster, England: MTP Press.

Chalupa, W. 1984. Discussion of protein symposium. J. Dairy Sci. 67:1134.

Chalupa, W., and G. C. Scott. 1976. Protein nutrition of growing cattle. P. 13 *in* Tracer Studies on Non-Protein Nitrogen for Ruminants. Vienna: International Atomic Energy Agency.

Chamberlain, D. G., and P. C. Thomas. 1976. Efficiency of bacterial protein synthesis in the rumen of sheep receiving a diet of sugar beet pulp and barley. J. Sci. Food Agric. 27:231.

Chamberlain, D. G., and P. C. Thomas. 1979. Ruminal nitrogen metabolism and the passage of amino acids to the duodenum in sheep receiving diets containing hay and concentrates in various proportions. J. Sci. Food Agric. 30:677.

Chamberlain, D. G., and P. C. Thomas. 1980a. Protein digestion in cows and sheep given silage diets. Proc. 3rd EAAP Symp. on Protein Metabolism and Nutrition. EAAP Publ. 27:422.

Chamberlain, D. G., and P. C. Thomas. 1980b. The effects of urea and artificial saliva on rumen bacterial protein synthesis in sheep receiving a high-cereal diet. J. Sci. Food Agric. 31:432.

Christensen, H. N. 1982. Interorgan amino acid nutrition. Physiol. Rev. 62:1193.

Church, D. C. 1969. Passage of digesta through the gastrointestinal tract. *In* Digestive Physiology and Nutrition of Ruminants, Vol. 1. Corvallis: Oregon State University Book Stores.

Clark, J. H. 1975a. Nitrogen metabolism in ruminants: Protein solubility and rumen bypass of protein and amino acids. P. 261 *in* Protein Nutritional Quality of Foods and Feeds, Vol. 1, Part 2, Mendel Friedman, ed. New York: Marcell Dekker, Inc.

Clark, J. H. 1975b. Lactational responses to postruminal administration of proteins and amino acids. J. Dairy Sci. 58:1178.

Clark, R. T. J. 1977. Methods for studying gut microbes. P. 1 *in* Microbial Ecology of the Gut, R. T. J. Clarke and T. Bauchop, eds. New York: Academic Press.

Clarke, E. M. W., G. M. Ellinger, and A. T. Phillipson. 1966. The influence of the diet on the nitrogenous components passing to the duodenum and through the lower ileum of sheep. Proc. R. Soc., Ser. B 166:63.

Cocimano, M. R., and R. A. Leng. 1967. Metabolism of urea in sheep. Br. J. Nutr. 21:353.

Coelho da Silva, J. F., R. C. Seeley, D. J. Thomson, D. E. Beever, and D. G. Armstrong. 1972a. The effect in sheep of physical form on the sites of digestion of a dried lucerne diet. 2. Sites of nitrogen digestion. Br. J. Nutr. 28:43.

Coelho da Silva, J. F., R. C. Seeley, D. E. Beever, J. H. D. Prescott, and D. G. Armstrong. 1972b. The effect in sheep of physical form and stage of growth on the sites of digestion of a dried grass. Br. J. Nutr. 28:357.

Cole, N. A., R. R. Johnson, and F. N. Owens. 1976a. Influence of roughage level and corn processing method on the site and extent of digestion by beef steers. J. Anim. Sci. 43:490.

Cole, N. A., R. R. Johnson, F. N. Owens, and J. R. Males. 1976b. Influence of roughage level and corn processing method on microbial protein synthesis by beef steers. J. Anim. Sci. 43:497.

Coleman, G. S. 1975. The interrelationship between rumen ciliate protozoa and bacteria. P. 149 *in* Digestion and Metabolism in the Ruminant, I. W. McDonald and A. C. I. Warner, eds. Armidale, NSW, Australia: The University of New England Publishing Unit.

Combe, E., E. Attaix, and M. Arnal. 1979. Protein turnover in the digestive tissues of the lamb throughout development. Ann. Rech. Vet. 10:436.

Conrad, H. R., J. W. Hibbs, and A. D. Pratt. 1960. Nitrogen metabolism in dairy cattle. I. Efficiency of nitrogen utilization by lactating cows fed various forages. Ohio Agric. Exp. Stn. Res. Bull. 861.

Cotta, M. A., and J. B. Russell. 1982. The effect of peptides and amino acids on efficiency of rumen bacterial protein synthesis in continuous culture. J. Dairy Sci. 65:226.

Cottrill, B. R., D. E. Beever, A. R. Austin, and D. F. Osbourn. 1982. The effect of protein and nonprotein nitrogen supplements to maize silage on total amino acid supply in young cattle. Br. J. Nutr. 48:527.

Coward, B. J., and P. J. Buttery. 1982. Metabolism of perfused ruminant muscle. J. Agric. Sci., Camb. 98:307.

Craig, W. M. 1981. Characteristics and effects of ruminal proteolysis on ruminant nutrition. Ph.D. dissertation. Texas A&M University, College Station.

Craig, W. M., and G. A. Broderick. 1984. Amino acids released during protein degradation by rumen microbes. J. Anim. Sci. 58:436.

Crawford, R. J., Jr., W. H. Hoover, C. J. Sniffen, and B. A. Crooker. 1978. Degradation of feedstuff nitrogen in the rumen *vs* nitrogen solubility in three solvents. J. Anim. Sci. 46:1768.

Crooker, B. A., C. J. Sniffen, W. H. Hoover, and L. L. Johnson. 1978. Solvents for soluble nitrogen measurements in feedstuffs. J. Dairy Sci. 61:437.

Danfaer, A. 1979. Quantitative protein and energy metabolism in a high-yielding dairy cow. *In* Protein Utilization in Farm Animals, Vol. I. Copenhagen: Inst. An. Sci., The Royal Vet. and Agric. Univ.

Danfaer, A., I. Thysen, and T. Ostergaard. 1980. The effect of the level of dietary protein on milk production. I. Milk yield, liveweight gain and health. Beretning fra Statens Husdyrbrugs forsøg 492. København.

Davis, S. R., T. N. Barry, and G. A. Hughson. 1981. Protein synthesis in tissues of growing lambs. Br. J. Nutr. 46:409.

Delfosse-Debusscher, J., D. Thines-Semoux, M. Vanbelle, and B. Latteur. 1979. Contribution of protozoa to the rumen cellulolytic activity. Ann. Rech. Vet. 10:255.

Demarquilly, C., J. Andrieu, D. Sauvant, and J. P. Dulphy. 1978. Composition et valeur nutritive des aliments (composition and nutritive value of feeds). P. 469 *in* Alimentation des Ruminants. Versailles: INRA Publications.

Demeyer, D. I., and C. J. Van Nevel. 1979. Effect of defaunation on the metabolism of rumen microorganisms. Br. J. Nutr. 42:515.

Down, P. F., L. Agostini, J. Murison, and O. M. Wrong. 1972. Interrelations of faecal ammonia, pH and bicarbonate: Evidence of colonic absorption of ammonia by non-ionic diffusion. Clin. Sci. 43:101.

Downes, A. M. 1961. On the amino acids essential for the tissues of the sheep. Aust. J. Biol. Sci. 14:254.

Driedger, A., and E. E. Hatfield. 1972. Influence of tannins on the nutritive value of soybean meal for ruminants. J. Anim. Sci. 34:465.

Dror, Y., and H. Tagari. 1980. Indirect measurement of metabolic and endogenic nitrogen. Nutr. Rep. Int. 21:721.

Ehle, F. R., M. R. Murphy, and J. H. Clark. 1982. *In situ* particle size reduction and the effect of particle size on degradation of crude protein and dry matter in the rumen of dairy steers. J. Dairy Sci. 65:963.

el-Shazly, K. 1958. Studies on the nutritive value of some common Egyptian feedingstuffs. I. Nitrogen retention and ruminal ammonia curves. J. Agric. Sci., Camb. 51:149.

Erdman, R. A. 1982. Methods for prediction of rumen protein degradation in the lactating cow. P. 39 *in* Proc. Md. Nutr. Conf.

Erfle, J. D., F. D. Sauer, and S. Mahadevan. 1977. Effect of ammonia concentration on activity of enzymes of ammonia assimilation and on synthesis of amino acid by mixed rumen bacteria in continuous culture. J. Dairy Sci. 60:1064.

Evans, E. 1981a. An evaluation of the relationships between dietary parameters and rumen liquid turnover rate. Can. J. Anim. Sci. 61:91.

Evans, E. 1981b. An evaluation between dietary parameters and rumen solid turnover rate. Can. J. Anim. Sci. 61:97.

Faichney, G. J. 1975. The use of markers to partition digestion within the gastrointestinal tract of ruminants. P. 277 *in* Digestion and Metabolism in the Ruminant, I. W. McDonald and A. C. I. Warner, eds. Armidale, NSW, Australia: The University of New England Publishing Unit.

Faichney, G. J. 1980. The use of markers to measure digesta flow from the stomach of sheep fed once daily. J. Agric. Sci. 94:313.

Faichney, G. J., D. E. Beever, and J. R. Black. 1980. A comparison of some recent approaches to the assessment of protein value in diets for ruminants. Proc. 3rd EAAP Symp. on Protein Metabolism and Nutrition. EAAP Publ. 27:432.

Fenderson, C. L., and W. G. Bergen. 1972. Effect of ration composition and protein level on plasma free trytophan and ruminal microbial tryptophan content in sheep. J. Anim. Sci. 35:896.

Ferguson, K. A. 1975. The protection of dietary proteins and amino acids against microbial fermentation in the rumen. P. 448 *in* Digestion and Metabolism in the Ruminant, I. W. McDonald and A. C. I. Warner, eds. Armidale, NSW, Australia: The University of New England Publishing Unit.

Ferguson, K. A., J. A. Hemsley, and P. J. Reis. 1967. Nutrition and wool growth. The effect of protecting dietary protein from microbial degradation. Aust. J. Sci. 30:215.

Ferrell, C. L. W. N. Garrett, and N. Hinman. 1976. Growth and development and composition of the udder and gravid uterus of beef heifers during pregnancy. J. Anim. Sci. 42:1477.

Forrest, W. W., and D. J. Walker. 1971. The generation and utilization of energy during growth. Adv. Microb. Physiol. 5:213.

Fox, D. G., R. G. Crickenberger, W. G. Bergen, and J. R. Black. 1976. A net protein system for formulating rations. J. Anim. Sci. 43:321.

Fox, D. G., C. J. Sniffen, and P. J. Van Soest. 1982. A net protein system for cattle: Meeting protein requirements of cattle. P. 280 *in* Protein Requirements for Cattle: Proceedings of an International Symposium, F. N. Owens, ed. MP-109. Stillwater, Okla.: Division of Agriculture, Oklahoma State University.

Ganev, G., E. R. Ørskov, and R. Smart. 1979. The effect of roughage on concentrate feeding and rumen retention time on total degradation of protein in the rumen. J. Agric. Sci., Camb. 93:651.

Geay, Y. 1980. Protein requirements of growing and fattening cattle. Proc. 3rd EAAP Symp. on Protein Metabolism and Nutrition. EAAP Pub. 27:803.

Geay, Y. 1984. Energy and protein utilization in growing cattle. J. Anim. Sci. 58:766.

Glenn, B. P., D. R. Waldo, H. F. Tyrrell, and H. K. Goering. 1983. Effect of increasing alfalfa and corn silage insolubilities on performance of cows in early lactation. J. Dairy Sci. 66 (Suppl. 1):146.

Goering, H. K. 1976. A laboratory assessment on the frequency of overheating in commercial dehydrated alfalfa samples. J. Anim. Sci. 43:869.

Goering, H. K., and P. J. Van Soest. 1972. Forage fiber analyses. USDA Agric. Handb. No. 379.

Goering, H. K., and D. R. Waldo. 1974. Processing effects on protein utilization by ruminants. P. 25 *in* Proc. Cornell Nutr. Conf.

Goering, H. K., and D. R. Waldo. 1978. The effects of dehydration on protein utilization in ruminants. *In* Proc. 2nd Int. Green Crop Drying Congress. Saskatoon, Canada: University of Saskatchewan.

Goering, H. K., C. H. Gordon, R. W. Hemken, D. R. Waldo, P. J. Van Soest, and L. W. Smith. 1972. Analytical estimates of nitrogen digestibility in heat damaged forages. J. Dairy Sci. 55:1275.

Goering, H. K., D. R. Waldo, and R. S. Adams. 1974. Nitrogen digestibility of wilted hay crop silages. P. 623 *in* Technique and Forage Conservation and Storage. Proc. XII Int. Grassl. Congr., Moscow.

Goetsch, A. L., and F. N. Owens. 1985. The effects of commercial processing method of cottonseed meal on site and extent of digestion in cattle. J. Anim. Sci. 60:803.

Gonzalez, J. S., J. J. Robinson, I. McHattie, and A. Z. Mehrez. 1979. The use of lactating ewes in evaluating protein sources for ruminants. Proc. Nutr. Soc. 38:145A.

Grummer, R. R., and J. H. Clark. 1982. Effect of dietary nitrogen solubility on lactation performance and protein and dry matter degradation *in situ*. J. Dairy Sci. 65:1432.

Hagemeister, V. H., and W. Kaufmann. 1974. Der einflub der rationsgestaltung auf die verfugbarkeit von protein-N bzw. Aminosaure-N im darm der milchkuh. Kiel. Milchwirtsch. Forschungsber. 26:199.

Hagemeister, V. H., and E. Pfeffer. 1973. Der einflub von formaldehydbehandeltem kasein und sojaschrot auf die mikrobiellen protein-umesetzungen in den vormagen und die aminosaureversorgung im darm der milchkuh. Z. Tierphysiol., Tierernaehr. Futtermittelkd. 31:275.

Hagemeister, H., W. Kaufmann, and E. Pfeffer. 1976. Factors influencing the supply of nitrogen and amino acids to the intestine of dairy cows. P. 425 *in* Protein Metabolism and Nutrition, D. J. A. Cole, L. N. Boorman, P. J. Buttery, D. Lewis, R. J. Neale, and H. Swan, eds. Boston: Butterworths.

Harmeyer, H., H. Holler, and H. Hartens. 1976. Estimate of microbial protein synthesis *in vitro* by the simultaneous use of three different isotopic markers. P. 69 *in* Tracer Studies in Non-Protein Nitrogen for Ruminants. Vienna: International Atomic Energy Agency.

Harris, C. I., and G. Milne. 1980. The urinary excretion of N-methyl histidine in sheep: An invalid index of muscle protein breakdown. Br. J. Nutr. 44:129.

Harris, C. I., and G. Milne. 1981. The excretion of N-methyl histidine by cattle: Validation as an index of muscle protein breakdown. Br. J. Nutr. 45:411.

Harris, L. E., L. C. Kearl, and P. V. Fonnesbeck. 1972. Use of regression equations in predicting availability of energy and protein. J. Anim. Sci. 35:658.

Harrison, D. G., and A. B. McAllan. 1980. Factors affecting microbial growth yields in the reticulo-rumen. P. 205 *in* Digestive Physiology and Metabolism in Ruminants, Y. Ruckebusch and P. Thivend, eds. Lancaster, England: MTP Press.

Harrison, D. G., D. E. Beever, D. J. Thompson, and D. F. Osbourn. 1975. Manipulation of rumen fermentation in sheep by increasing the rate of flow of water from the rumen. J. Agric. Sci., Camb. 85:93.

Harrison, D. G., D. E. Beever, D. J. Thompson, and D. F. Osbourn. 1976. Manipulation of fermentation in the rumen. J. Sci. Food Agric. 27:617.

Hartnell, G. F., and L. D. Satter. 1979. Determination of rumen fill, retention time and ruminal turnover rates of ingesta at different stages of lactation in dairy cows. J. Anim. Sci. 48:381.

Haselbach, C. 1980. Zur Proteinversorgung der Wiederkauer. P. 16 *in* Proteinversorgung Landwirtschaftlicher Nutztiere. Zurich: Eidgenossische Technishe Hochschule.

Hecker, J. F. 1971. Metabolism of nitrogenous compounds in the large intestine of sheep. Br. J. Nutr. 25:85.

Hegsted, D. M., and Y. Chang. 1965. Protein utilization in growing rats. I. Relative growth index as a bioassay procedure. J. Nutr. 85:159.

Heitmann, R. N., and E. N. Bergman. 1980. Integration of amino acid metabolism in sheep: Effects of fasting and acidosis. Am. J. Physiol. 239:E248.

Hemsley, J. A. 1975. Effect of high intake of sodium chloride on the utilization of a protein concentrate by sheep. Aust. J. Agric. Res. 26:709.

Henderickx, H., and J. Martin. 1963. *In vitro* study of the nitrogen metabolism in the rumen. C. R. Rech. Inst. Rech. Sci. Ind. Agric. Bruxelles 31:7.

Hespell, R. B., and M. P. Bryant. 1979. Efficiency of rumen microbial growth: Influence of some theoretical and experimental factors on Y_{ATP}. J. Anim. Sci. 49:1640.

Hibberd, C. A. 1982. Varietal, environmental and processing effects on the nutritive characteristics of sorghum grain. Ph.D. thesis. Oklahoma State University, Stillwater.

Hill, K. J., and J. L. Mangan. 1964. The formation and distribution of methylamine in the ruminant digestive tract. Biochem. J. 93:39.

Hogan, J. P. 1973. Intestinal digestion of subterranean clover by sheep. Aust. J. Agric. Res. 24:587.

Hogan, J. P. 1975. Quantitative aspects of nitrogen utilization in ruminants. J. Dairy Sci. 58:1164.

Hogan, J. P., and R. H. Weston. 1967. The digestion of two diets of differing protein content but with similar capacities to sustain wool growth. Aust. J. Agric. Res. 18:973.

Hogan, J. P., and R. H. Weston. 1970. Quantitative aspects of microbial protein synthesis in the rumen. P. 474 *in* Physiology of Digestion and Metabolism in the Ruminant, A. T. Phillipson, ed. Newcastle-upon-Tyne: Oriel Press.

Holter, J. A., and J. T. Reid. 1959. Relationship between the concentrations of crude protein and apparently digestible protein in forages. J. Anim. Sci. 18:1339.

Hoogenraad, N. J., and F. J. R. Hird. 1970. The chemical composition of rumen bacteria and cell walls from rumen bacteria. Br. J. Nutr. 24:119.

Hoover, W. H. 1978. Digestion and absorption in the hindgut of ruminants. J. Anim. Sci. 46:1789.

Hoover, W. H., and R. N. Heitmann. 1975. Cecal nitrogen metabolism and amino acid absorption in the rabbit. J. Nutr. 105:245.

Huber, J. T., and L. Kung, Jr. 1981. Protein and nonprotein nitrogen utilization in dairy cattle. J. Dairy Sci. 64:1170.

Hume, I. D. 1970. Synthesis of microbial protein in the rumen. III. The effect of dietary protein. Aust. J. Agric. Res. 21:305.

Hume, I. D. 1974. The proportion of dietary protein escaping degradation in the rumen of sheep fed on various protein concentrates. Aust. J. Agric. Res. 25:155.

Hume, I. D., and P. R. Bird. 1969. Synthesis of microbial protein in the rumen. IV. The influence of the level and form of dietary sulphur. Aust. J. Agric. Res. 21:297.

Hume, I. D., and D. B. Purser. 1974. Ruminal and post-ruminal protein digestion in sheep fed on subterranean clover harvested at four stages of maturity. Aust. J. Agric. Res. 26:199.

Hume, I. D., D. R. Jacobson, and G. E. Mitchell. 1972. Quantitative studies on amino acid absorption in sheep. J. Nutr. 102:495.

Hungate, R. E. 1966. The Rumen and Its Microbes. New York: Academic Press.

Hvelplund, T., and P. D. Möller. 1976. Nitrogen metabolism in the gastrointestinal tract of cows fed silage. Z. Tierphysiol., Tierernaehr. Futtermittelkd. 37:183.

Ibrahim, E. A., and J. R. Ingalls. 1971. Microbial protein biosynthesis in the rumen. J. Dairy Sci. 55:971.

Institut National de la Recherche Agronomique. 1978. Alimentation des Ruminants. Versailles: INRA Publications.

Isichei, C. O. 1980. The Role of Monensin in Protein Metabolism in Steers. Ph.D. thesis. Michigan State University, East Lansing.

Isichei, C. O., and W. G. Bergen. 1980. The effect of monensin on the composition of abomasal nitrogen flow in steers fed grain and silage rations. J. Anim. Sci. 51:Suppl.1:371.

Johns, J. T., and W. C. Bergen. 1973. Studies on amino acid uptake by ovine small intestine. J. Nutr. 103:1581.

Johnson, D. E., and W. C. Bergen. 1982. Fraction of organic matter digestion occurring in the rumen: A partial literature summary. P. 113 *in* Protein Requirements for Cattle: Proceedings of an International Symposium, F. N. Owens, ed. MP-109. Stillwater, Okla.: Division of Agriculture, Oklahoma State University.

Journet, M., and R. Vérité. 1979. Predicting equations of duodenal flow in dairy cattle effects of level of feeding and proportion of concentrate in the diet. Ann. Rech. Vet. 10:303.

Kang-Meznarich, J. H., and G. A. Broderick. 1981. Effects of incremental urea supplementation on ruminal ammonia concentration and bacterial protein formation. J. Anim. Sci. 51:422.

Kaufmann, W. 1977a. Economic and other considerations governing decisions on the advisability of incorporating additional and new sources of protein and non-protein nitrogen into the diet of beef cattle (including fattening cattle). P. 101 *in* Protein and Non-Protein Nitrogen for Ruminants. Recent Developments in the Use of New Sources. Elmsford, N.Y.: Pergamon Press.

Kaufmann, W. 1977b. Calculation of the protein requirements for dairy cows according to measurements of N metabolism. Proc. 2nd Int. Symp. on Protein Metabolism and Nutrition. EAAP Pub. 22:130.

Kaufmann, W. 1979. Protein digestibility and the factorial calculation of crude protein requirements of dairy cows. Z. Tierphysiol., Tierernaehr. Futtermittelkund. 41:326.

Kennedy, P. M., and L. P. Milligan. 1978. Transfer of urea from the blood to the rumen of sheep. Br. J. Nutr. 40:149.

Kennedy, P. M., and L. P. Milligan. 1980. The degradation and utilization of endogenous urea in the gastro intestinal tract of ruminants. A review. Can. J. Anim. Sci. 60:205.

Kennedy, P. M., R. J. Christopherson, and L. P. Milligan. 1976. The effect of cold exposure of sheep on digestion, rumen turnover time and efficiency of microbial synthesis. Br. J. Nutr. 36:231.

Kennedy, P. M., R. T. J. Clarke, and L. P. Milligan. 1981. Influences of dietary sucrose and urea on transfer of endogenous urea to the rumen of sheep and numbers of piteleal bacteria. Br. J. Nutr. 46:533.

Kennedy, P. M., R. J. Christopherson, and L. P. Milligan. 1982. Effects of cold exposure on feed protein degradation, microbial protein synthesis and transfer of plasma urea to the rumen of sheep. Br. J. Nutr. 47:521.

Kern, D. L., L. L. Slyter, E. C. Leffel, J. M. Weaver, and R. R. Oltjen. 1974. Ponies vs steers: Microbial and chemical characteristics of intestinal digesta. J. Anim. Sci. 38:559.

Klopfenstein, T. W. 1980. Protein degradation in ruminants. Proc. Fla. Nutr. Conf., Gainesville, Fla.

Klopfenstein, T., R. Britten, and R. Stock. 1982. Nebraska growth system. P. 310 *in* Protein Requirements for Cattle: Proceedings of an International Symposium. F. N. Owens, ed. MP-109. Stillwater, Okla.: Division of Agriculture, Oklahoma State University.

Krishnamoorthy, U., T. V. Muscato, C. J. Sniffen, and P. J. Van Soest. 1982. Nitrogen fractions in selected feedstuffs. J. Dairy Sci. 65:217.

Krishnamoorthy, U., C. J. Sniffen, M. D. Stern, and P. J. Van Soest. 1983. Evaluation of a rumen dynamic mathematical model and *in vitro* simulated rumen proteolysis to estimate rumen escape nitrogen in feedstuffs. Br. J. Nutr. 50:555.

Kristensen, E. S., P. D. Moller, and T. Hvelplund. 1982. Estimation of the effective protein degradability in the rumen of cows using the nylon bag technique combined with outflow rate. Acta Agric. Scand. 32:123.

Kropp, J. R., R. R. Johnson, J. R. Males, and F. N. Owens. 1977a. Microbial protein synthesis with low quality roughage rations: Isonitrogenous substitution of urea for soybean meal. J. Anim. Sci. 45:837.

Kropp, J. R., R. R. Johnson, J. R. Males, and F. N. Owens. 1977b. Microbial protein synthesis with low quality roughage rations: Level and source of nitrogen. J. Anim. Sci. 46:844.

Landis, J. 1979. Die protein-und energieversorgung der Milchkuh. Schweiz. Landwirtsch. Monatsh. 57:381.

Lane, R. J., and T. L. Cross. 1985. Spreadsheet applications for Animal Nutrition and Feeding. Reston, Va.: Reston Publ. Co.

Laughren, L. C., and A. W. Young. 1979. Duodenal nitrogen flow in response to increasing dietary crude protein in sheep. J. Anim. Sci. 49:211.

Leibholz, J., and P. E. Hartmann. 1972. Nitrogen metabolism in sheep. I. The effect of protein and energy intake on the flow of digesta into the duodenum and on the digestion and absorption of nutrients. Aust. J. Agric. Res. 23:1059.

Leibholz, J., and R. C. Kellaway. 1979. Amino acid requirements for microbial protein synthesis. Ann. Rech. Vet. 10:274.

Lemons, J. A., E. W. Adcock III, M. D. Jones, Jr., M. A. Naughton, G. Meschia, and F. C. Battaglia. 1976. Umbilical uptake of amino acids in the unstressed fetal lamb. J. Clin. Invest. 58:1428.

Leng, R. A., and J. V. Nolan. 1984. Nitrogen metabolism in the rumen. J. Dairy Sci. 67:1072.

Leng, R. A., T. J. Kempton, and J. V. Nolan. 1977. Non-protein nitrogen and bypass proteins in ruminant diets. Aust. Meat Res. Committee Rev. 33:1.

Lewis, D. 1962. The inter-relationships of individual proteins and carbohydrates during fermentation in the rumen of the sheep. II. The fermentation of starch in the presence of proteins and other substances containing nitrogen. J. Agric. Sci., Camb. 58:73.

Lindberg, J. E. 1982. Ruminal flow rate of soya-bean meal, rapeseed meal and cottonseed meal in cows fed at maintenance and at three times maintenance. J. Agric. Sci., Camb. 98:689.

Ling, J. R., and P. J. Buttery. 1977. The simultaneous use of ribonucleic acid, ^{35}S, 2,6-diaminopimelic acid and 2-aminoethylphosphoric acid as markers of microbial nitrogen entering the duodenum of sheep. Br. J. Nutr. 39:165.

Little, C. O., W. Burroughs, and W. Woods. 1963. Nutritional significance of soluble nitrogen in dietary proteins for ruminants. J. Anim. Sci. 22:358.

Lobley, G. E., V. Milne, J. M. Lovie, P. J. Reeds, and K. Pennie. 1980. Whole body and tissue protein synthesis in cattle. Br. J. Nutr. 43:491.

Loerch, S. C., L. L. Berger, S. D. Plegge, and G. C. Fahey, Jr. 1983. Digestibility and rumen escape of soybean meal, blood meal, meat and bone meal and dehydrated alfalfa nitrogen. J. Anim. Sci. 57:1037.

Lu, C. D., N. A. Jorgensen, R. J. Early, and K. B. Smith. 1981. Flow of nitrogenous compounds in the gastro-intestinal tract of sheep fed alfalfa protein concentrate prepared by various methods. P. 415 in Am. Soc. Anim. Sci. (Abstr.).

Madsen, J., P. D. Moller, K. V. Thomsen, and T. Hvelplund. 1977. The influence of chemical treated feed protein on the nitrogen metabolism of the ruminant. Beretning fra Statens Husdyrbrugs forsøg 458. København.

Maeng, W. J., and R. L. Baldwin. 1975. Factors influencing rumen microbial growth rates and yields: Effects of urea and amino acids over time. J. Dairy Sci. 59:643.

Maeng, W. J., C. J. Van Nevel, R. L. Baldwin, and J. G. Morris. 1975. Rumen microbial growth rates and yields: Effects of amino acids and protein. J. Dairy Sci. 59:68.

Mahadevan, S., J. D. Erfle, and F. D. Sauer. 1980. Degradation of soluble and insoluble proteins by *Bacteroides amylophilus* protease and by rumen microorganisms. J. Anim. Sci. 50:723.

Mangan, J. L. 1972. Quantitative studies on nitrogen metabolism in the bovine rumen. The rate of proteolysis of casein and ovalbumin and the release and metabolism of free amino acids. Br. J. Nutr. 27:261.

Mason, V. C. 1969. Some observations on the distribution and origin of nitrogen in sheep faeces. J. Agric. Sci., Camb. 73:99.

Mason, V. C. 1981. Factors influencing fecal nitrogen excretion in sheep. 1. The digestibility and level of intake of pelleted diets. Z. Tierphysiol., Tierernaehr. Futtermittelkd. 45:161.

Mason, V. C., and J. H. Fredericksen. 1979. Partition of the nitrogen in sheep faeces with detergent solutions, and its application to the estimation of the true digestibility of dietary nitrogen and the excretion of non dietary faecal nitrogen. Z. Tierphysiol., Tierernaehr. Futtermittelkd. 41:121.

Mason, V. C., and F. White. 1971. The digestion of bacterial mucopeptide constituents in the sheep. 1. The metabolism of 2,6-diaminopimelic acid. J. Agric. Sci., Camb. 77:91.

Mason, V. C., M. P. Narang, J. C. Ononiwu, and P. Kessank. 1977. The relationship between nitrogen metabolism in the hind-gut and nitrogen excretion. In 2nd Int. Symp. on Protein Metabolism and Nutrition. Wageningen: Centre for Agricultural Publishing and Documentation.

Mason, V. C., P. Kessank, J. C. Ononiwu, and M. P. Narang. 1981. Factors influencing faecal nitrogen excretion in sheep. 2. Carbohydrate fermentation in the caecum and large intestine. Z. Tierphysiol., Tierernaehr. Futtermittelkd. 45:174.

Mathers, J. C., and E. L. Miller. 1979. The determination of amino acid requirements. P. 3.1 in Protein Metabolism in the Ruminant, P. J. Buttery, ed. London: Agricultural Research Council.

Mathers, J. C., and E. L. Miller. 1981. Quantitative studies of food protein degradation and the energetic efficiency of microbial protein synthesis in the rumen of sheep given chopped lucerne and rolled barley. Br. J. Nutr. 45:587.

Mathison, G. W., and L. P. Milligan. 1971. Nitrogen metabolism in sheep. Br. J. Nutr. 25:351.

Matis, J. E., and H. D. Tolley. 1980. On the stochastic modeling of tracer kinetics. Fed. Proc. 39:104.

Matthews, D. M. 1972. Intestinal absorption of amino acids and peptides. Proc. Nutr. Soc. 31:171.

Mazanov, A., and J. V. Nolan. 1976. Simulation of the dynamics of nitrogen metabolism in sheep. Br. J. Nutr. 35:149.

MacRae, J. C. 1978. Nitrogen transactions in the gut and tissues. P. 6.1 in Ruminant Digestion and Feed Evaluation, D. F. Osbourn, D. E. Beever, and D. J. Thompson, eds. London: Agricultural Research Council.

MacRae, J. C., and M. J. Ulyatt. 1974. Quantitative digestion of fresh herbage by sheep. J. Agric. Sci., Camb. 82:309.

MacRae, J. C., M. J. Ulyatt, P. D. Pearce, and J. Hendtlass. 1972. Quantitative intestinal digestin of nitrogen in sheep given formaldehyde-treated and untreated casein supplements. Br. J. Nutr. 27:39.

McAllan, A. B., and R. H. Smith. 1983. Estimation of flows of organic matter and nitrogen components in postruminal digesta and effects of level of dietary intake and physical form of protein supplement on such estimates. Br. J. Nutr. 49:119.

McCormick, M. E., and K. E. Webb, Jr. 1982. Plasma free, erythrocyte free and plasma peptide amino acid exchange of calves in steady state and fasting metabolism. J. Nutr. 112:276.

McDonald, I. 1981. A revised model for the estimation of protein degradability in the rumen. J. Agric. Sci., Camb. 96:251.

McDonald, I. W. 1952. The role of ammonia in ruminal digestion of protein. Biochem. J. 51:86.

McDonald, I. W. 1954. The extent of conversion of food protein to microbial protein in the rumen of the sheep. Biochem. J. 56:120.

McLaren, G. A., G. C. Anderson, J. A. Welch, C. O. Campbell, and G. S. Smith. 1960. Diethylstilbestrol and length of preliminary period in the utilization of crude biuret and urea by lambs. II. Various aspects of nitrogen metabolism. J. Anim. Sci. 19:44.

McMeniman, N. P. 1976. Ph.D. thesis. University of Newcastle-upon-Tyne, England. Cited in ARC (1980).

McMeniman, N. P., D. Ben-Ghedalia, and D. G. Armstrong. 1976. Nitrogen-energy interactions in rumen fermentation. Page 217 *in* Protein Metabolism and Nutrition. D. J. A. Cole, K. N. Boorman, P. J. Buttery, D. Lewis, R. J. Neale, and H. Swan, eds. Boston: Butterworths.

Meers, J. L. 1973. Growth of bacteria in mixed culture. CRC Critical Reviews in Microbiology 2:139.

Mehrez, A. Z., and E. R. Ørskov. 1977. A study of the artificical fibre bag technique for determining the digestibility of feeds in the rumen. J. Agric. Sci., Camb. 88:645.

Mehrez, A. Z., E. R. Ørskov, and I. McDonald. 1977. Rates of rumen fermentation in relation to ammonia concentration. Br. J. Nutr. 38:447.

Mehrez, A. Z., E. R. Ørskov, and J. Opstvedt. 1980. Processing factors affecting degradability of fish meal in the rumen. J. Anim. Sci. 50:737.

Meier, P., C. Teng, F. C. Battaglia, and C. Meschia. 1981. The rate of amino acid nitrogen and total nitrogen accumulation in the fetal lamb. Proc. Soc. Exp. Biol. Med. 167:463.

Mepham, T. F. 1982. Amino acid utilization by lactating mammary gland. J. Dairy Sci. 65:287.

Mercer, J. R., S. A. Allen, and E. L. Miller. 1980. Rumen bacterial protein synthesis and the proportion of dietary protein escaping degradation in the rumen of sheep. Br. J. Nutr. 43:421.

Merchen, N. R. 1981. Effect of conservation method on digestion of alfalfa by ruminants. Ph.D. thesis. University of Wisconsin, Madison.

Merchen, N. R., and L. D. Satter. 1983a. Digestion of nitrogen by lambs fed alfalfa conserved as baled hay or as low moisture silage. J. Anim. Sci. 56:943.

Merchen, N. R., and L. D. Satter. 1983b. Changes in nitrogenous compounds and sites of digestion of alfalfa harvested at different moisture contents. J. Dairy Sci. 66:789.

Merchen, N., T. Hanson, and T. Klopfenstein. 1979. Ruminal bypass of brewers dried grains protein. J. Anim. Sci. 49:192.

Mertens, D. R., and L. O. Ely. 1979. A dynamic model of fiber digestion and passage in the ruminant for evaluating forage quality. J. Anim. Sci. 49:1085.

Miller, E. L. 1973. Evaluation of foods as sources of nitrogen and amino acids. Proc. Nutr. Soc. 32:79.

Miller, E. L. 1980. Protein value of feedstuffs for ruminants. Volume 3. Vicia faba. Feeding Value, Processing and Viruses *in* World Crops: Production, Utilization, and Description, A. Bond, ed. Brussels: Marinus Nijhoff.

Minato, H. A. Endo, M. Higushi, Y. Octomo, and T. Vermara. 1966. Ecological treatise on rumen fermentation. 1. The fractionation of bacteria attached to the rumen digesta solids. J. Gen. Appl. Microbiol. 12:39.

Mitchell, H. H. 1962. Comparative Nutrition of Man and Domestic Animals, Vol. 1. New York: Academic Press.

Mohamed, O. E., and R. H. Smith. 1977. Measurements of protein degradation in the rumen. Proc. Nutr. Soc. 36:152A.

Möller, P. D., and T. Hvelplund. 1982. Nitrogen metabolism in the forestomachs of cows fed ammonia treated barley straw supplemented with increasing amounts of urea or soya bean meal. Z. Tierphysiol., Tierernaehr. Futtermittelkd. 48:46.

Möller, P. D., and K. V. Thomsen. 1977. Nitrogen utilization from forages by ruminants. Laboratorial and physiological measurements. P. 27 *in* Quality of Forage. Proc. of a seminar organized by NJF. Inst. for Husdjurens Utfodring och vard. Rapport nr 54.

Morrison, F. B. 1959. Feeds and Feeding. Clinton, Iowa: Morrison Publishing Co.

Mukherjee, R., and N. D. Kehar. 1949. Studies on protein metabolism. III. An improved technique for estimating the metabolic fecal nitrogen of cattle. Indian J. Vet. Sci. Anim. Husb. 19:99.

Munck, B. G. 1976. Amino acid transport. P. 73 *in* Protein Metabolism and Nutrition, D. J. A. Cole, K. N. Boorman, P. J. Buttery, D. Lewis, and R. J. Neale, eds. Boston: Butterworths.

National Research Council. 1976. Nutrient Requirements of Beef Cattle, 5th rev. ed. Washington, D.C.: National Academy of Sciences.

National Research Council. 1978. Nutrient Requirements of Dairy Cattle. 5th rev. ed. Washington, D.C.: National Academy of Sciences.

National Research Council. 1981. Nutritional Energetics of Domestic Animals and Glossary of Energy Terms. Washington, D.C.: National Academy Press.

National Research Council. 1982. United States-Canadian Tables of Feed Composition. Washington, D.C.: National Academy Press.

National Research Council. 1984. Nutrient Requirements of Beef Cattle. 6th rev. ed. Washington, D.C.: National Academy Press.

Netemeyer, D. T., L. J. Bush, and F. N. Owens. 1980. Effect of particle size of soybean meal on protein utilization in steers and lactating cows. J. Dairy Sci. 63:574.

Nikolic, J. A., and R. Filipovic. 1981. Degradation of maize protein in rumen contents. Influence of ammonia concentration. Br. J. Nutr. 45:111.

Nimrick, K. E., E. E. Hatfield, J. Kaminskil, and F. N. Owens. 1970. Quantitative assessment of supplemental amino acid needs for growing lambs fed urea as the sole nitrogen source. J. Nutr. 100:1293.

Nocek, J. E., K. A. Cummins, and C. E. Polan. 1979. Ruminal disappearance of crude protein and dry matter in feeds and combined effects of formulated rations. J. Dairy Sci. 62:1587.

Nolan, J. V. 1975. Quantitative models of nitrogen metabolism in sheep. P. 416 *in* Digestion and Metabolism in the Ruminant, I. W. McDonald and A. C. I. Warner, eds. Armidale, NSW, Australia: The University of New England Publishing Unit.

Nolan, J. V., and R. A. Leng. 1972. Dynamic aspects of ammonia and urea metabolism in sheep. Br. J. Nutr. 27:177.

Nolan, J. V., W. B. Norton, and R. A. Leng. 1972. Dynamic aspects of nitrogen metabolism in sheep. P. 13 *in* Tracer Studies on Non-Protein Nitrogen for Ruminants. Vienna: International Atomic Energy Agency.

Nolan, J. V., W. B. Norton, and R. A. Leng. 1976. Further studies of the dynamics of nitrogen metabolism in sheep. Br. J. Nutr. 35:127.

Nugent, J. H. A., and J. L. Mangan. 1978. Rumen proteolysis of fraction I leaf protein, casein and bovine serum albumin. Proc. Nutr. Soc. 37:48A.

Nugent, J. H. A., and J. L. Mangan. 1981. Characteristics of the rumen proteolysis of fraction I (18S) leaf protein from lucerne (*Medicago sativa* L.). Br. J. Nutr. 46:39.

Oldham, J. D. 1980. Amino acid requirements for lactation in high yielding dairy cows. P. 33 *in* Proc. Nottingham Univ. Conf. for Feed Manufacturers.

Oldham, J. D. 1981. Amino acid requirements for lactation in high yielding dairy cows. P. 33, *in* Recent Advances in Animal Nutrition, W. Haresign, ed. London: Butterworths.

Oldham, J. D. 1984. Protein-energy interrelationships in dairy cattle. J. Dairy Sci. 67:1090.

Oldham, J. D., and G. Alderman. 1982. Recent advances in understanding protein energy interrelationships in intermediary metabolism in ruminants. P. 33 *in* Protein and Energy Supply for High Production of Milk and Meat. Oxford: Pergamon Press.

Oldham, J. D., and S. Tamminga. 1980. Amino acid utilization by dairy cows. I. Methods of varying amino acid supply. Livestock Prod. Sci. 7:437.

Oldham, J. D., J. D. Sutton, and A. B. McAllan. 1979. Protein digestion and utilization by dairy cows. Ann. Rech. Vet. 10:290.

Ørskov, E. R. 1982. Protein Nutrition in Ruminants. New York: Academic Press.

Ørskov, E. R., and I. McDonald. 1979. The estimation of protein degradability in the rumen from incubation measurements weighted according to rate of passage. J. Agric. Sci., Camb. 92:499.

Ørskov, E. R., and N. A. MacLeod. 1982. The determination of the minimal nitrogen excretion in steers and dairy cows and its physiological and practical implications. Br. J. Nutr. 47:625.

Ørskov, E. R., and A. Z. Mehrez. 1977. Estimation of extent of protein degradation from basal feeds in the rumen of sheep. Proc. Nutr. Soc. 36:78A.

Ørskov, E. R., C. Fraser, and I. McDonald. 1971a. Digestion of concentrates in sheep. 2. The effect of urea or fish-meal supplementation of barley diets on the apparent digestion of protein, fat, starch, and ash in the rumen, small intestine and the large intestine and calculation of volatile fatty acid production. Br. J. Nutr. 25:243.

Ørskov, E. R., C. Fraser, and I. McDonald. 1971b. Digestion of concentrates in sheep. 3. Effects of rumen fermentation of barley and maize diets on protein digestion. Br. J. Nutr. 26:477.

Ørskov, E. R., C. Fraser, and I. McDonald. 1972. Digestion of concentrates in sheep. 4. The effects of urea on digestion, nitrogen retention and growth in young lambs. Br. J. Nutr. 27:491.

Ørskov, E. R., C. Fraser, I. McDonald, and R. I. Smart. 1974. Digestion of concentrates in sheep. 5. The effect of adding fish meal and urea together to cereal diets on protein digestion and utilization by young sheep. Br. J. Nutr. 31:89.

Ørskov, E. R., F. D. DeB. Hovell, and F. Mould. 1980. The use of the nylon bag technique for evaluation of feedstuffs. Trop. Anim. Prod. 5:195.

Ørskov, E. R., M. Hughes-Jones, and M. E. Elimam. 1983. Studies on degradation and outflow rate of protein supplements in the rumen of sheep and cattle. Livestock Prod. Sci. 10:17.

Overman, O. R., O. F. Garrett, K. E. Wright, and F. P. Sanmann. 1939. Composition of milk of Brown Swiss cows, with summary of data on the composition of milk from cows of other dairy breeds. Ill. Agric. Exp. Stn. Bull. 457:575.

Owens, F. N., and W. G. Bergen. 1983. Nitrogen metabolism of ruminant animals. Historical perspective, current understanding and future implications. J. Anim. Sci. 57(Suppl. 2):498.

Owens, F. N., and R. A. Zinn. 1982. The standard reference system of protein bypass estimation. P. 352 in Protein Requirements for Cattle: Proceedings of an International Symposium, F. N. Owens, ed. MP-109. Stillwater, Okla.: Division of Agriculture, Oklahoma State University.

Oyaert, W., and J. H. Bouckaert. 1960. Quantitative aspects of food digestion in the rumen. Zentralbl. Veterinaermed. 7:929.

Paster, B. J., and E. Canale-Parola. 1982. Physiological diversity of rumen spirochetes. Appl. Environ. Microbiol. 43:686.

Pena, F., and L. D. Satter. 1983. Site and extent of digestion of solvent extracted and expeller processed cottonseed meals in lactating Holstein cows. J. Anim. Sci. 57(Suppl. 1):459.

Pena, F., H. Tagari, and L. D. Satter. 1983. Effect of heat treating whole cottonseed on rumen degradation and flow of protein to the small intestine in Holstein cows. J. Dairy Sci. 66(Suppl. 1):200.

Pena, F., H. Tagari, and L. D. Satter. 1985. The effect of heat treatment of whole cottonseeds on site and extent of protein digestion in dairy cows. J. Anim. Sci. (in press).

Phillips, W. A., K. E. Webb, and J. P. Fontenot. 1976. In vitro absorption of amino acids by the small intestine of sheep. J. Anim. Sci. 42:201.

Phillipson, A. T. 1964. The digestion and absorption of nitrogenous compounds in the ruminant. P. 71 in Mammalian Protein Metabolism I, H. N. Munro and J. B. Allison, eds. New York: Academic Press.

Pichard, G., and P. J. Van Soest. 1977. Protein solubility of ruminant feeds. P. 91 in Proc. Cornell Nutr. Conf.

Pilgrim, A. F., F. V. Gray, R. A. Weller, and C. B. Belling. 1970. Synthesis of microbial protein in the sheep's rumen and the proportion of dietary nitrogen converted into microbial nitrogen. Br. J. Nutr. 24:589.

Pirt, S. J. 1965. The maintenance energy of bacteria in growing cultures. Proc. R. Soc. B. 163:224.

Pittman, K. A., and M. P. Bryant. 1964. Peptides and other nitrogen sources for growth of Bacteroides ruminicola. J. Bacteriol. 88:401.

Plouzek, C. A., and A. Trenkle. 1982. Nitrogen fractions in feces of cattle. J. Anim. Sci. 55(Suppl. 1):451.

Poos, M. I., L. S. Bull, and R. W. Hemken. 1979a. Supplementation of diets with positive and negative urea fermentation potential using urea or soybean meal. J. Anim. Sci. 49:1417.

Poos, M. I., T. L. Hanson, and T. J. Klopfenstein. 1979b. Monensin effects on diet digestibility, ruminal protein bypass and microbial protein synthesis. J. Anim. Sci. 48:1516.

Poos, M., T. Klopfenstein, R. A. Britton, and D. G. Olson. 1980a. An enzymatic technique for determining ruminal protein degradation. J. Dairy Sci. 63(Suppl. 1):142.

Poos, M., T. Klopfenstein, R. A. Britton, and D. G. Olson. 1980b. A comparison of laboratory techniques to predict ruminal degradation of protein supplements. J. Anim. Sci. 51(Suppl. 1):389.

Potter, G. D., J. W. McNeill, and J. K. Riggs. 1971. Utilization of processed sorghum grain proteins by steers. J. Anim. Sci. 32:540.

Prange, R. W. 1981. Kinetics of digesta passage in lactating dairy cows. Ph.D. thesis. Univ. Wisconsin, Madison.

Preston, R. L. 1982. Empirical value of crude protein systems for feedlot cattle. P. 201 in Protein Requirements of Cattle: Proceedings of an International Symposium, F. N. Owens, ed. MP-109. Stillwater, Okla.: Division of Agriculture, Oklahoma State University.

Prigge, E. C., M. L. Galyean, F. N. Owens, D. G. Wagner, and R. R. Johnson. 1978. Microbial protein synthesis in steers fed processed corn rations. J. Anim. Sci. 46:249.

Rattray, P. V., W. N. Garrett, N. E. East, and N. Hinman. 1974. Growth, development, and composition of the ovine conceptus and mammary gland during pregancy. J. Anim. Sci. 38:613.

Razzaque, M. A., J. II. Topps, R. N. B. Kay, and J. M. Brockway. 1981. Metabolism of the nucleic acids of rumen bacteria by preruminant and ruminant lambs. Br. J. Nutr. 45:517.

Redman, R. G., R. C. Kellaway, and J. Leibholz. 1980. Utilization of low quality roughages: Effects of urea and protein supplements of differing solubility on digesta flows, intake and growth rate of cattle eating oaten chaff. Br. J. Nutr. 44:343.

Reid, J. T., and J. Robb. 1971. Relationship of body composition to energy intake and energetic efficiency. J. Dairy Sci. 54:553.

Reis, P. J., D. A. Tunks, and A. M. Downes. 1973. The influence of abomasal and intravenous supplements of sulfur-containing amino acids on wool growth rate. Aust. J. Biol. Sci. 26:249.

Reitnour, C. M., and R. L. Salsbury. 1972. Digestion and utilization of cecally infused protein by the equine. J. Anim. Sci. 35:1190.

Richardson, C. R. and E. E. Hatfield. 1978. The limiting amino acids in growing cattle. J. Anim. Sci. 26:740.

Robelin, J., and R. Daenicke. 1980. Variations of net requirements for cattle growth with liveweight, liveweight gain, breed and sex. Ann. Zootech. 29(n hors serie):99.

Robinson, P. H. 1983. Development and initial testing of an in vivo system to estimate rumen and whole tract digestion in lactating dairy cows. Ph.D. thesis. Cornell University.

Rode, L. M., D. C. Weakley, and L. D. Satter. 1985. Effect of forage amount and particle size in diets of lactating dairy cows on site of digestion and microbial protein synthesis. Can. J. Anim. Sci. 65:101.

Roels, J. A. 1980. Application of macroscopic principles to microbial metabolism. Biotech. Bioeng. 22:2457.

Roffler, R. E., and L. D. Satter. 1975a. Relationship between ruminal ammonia and nonprotein nitrogen utilization by ruminants. I. Development of a model for predicting nonprotein nitrogen utilization by cattle. J. Dairy Sci. 58:1880.

Roffler, R. E., and L. D. Satter. 1975b. Relationship between ruminal ammonia and nonprotein nitrogen utilization by ruminants. II. Application of published evidence to the development of a theoretical model for predicting nonprotein nitrogen utilization. J. Dairy Sci. 58:1889.

Rogers, J. A., B. C. Marks, C. L. Davis, and D. H. Clark. 1979. Alteration of rumen fermentation in steers by increasing rumen fluid dilution rate with mineral salts. J. Dairy Sci. 62:1599.

Rohr, K., M. Brandt, O. Castrillo, P. Lebzien, and G. Assmus. 1979. Der Einfluss eines teilweisen Ersatzes von Futterprotein durch Harnstoff auf den Stickstoff und Aminosaurenfluss am Duodenum. Landbauforsch. Volkenrode 29:32.

Rook, J. A., I. M. Brookes, and D. G. Armstrong. 1983. The digestion of untreated and formaldehyde-treated soya-bean and rapeseed meals by cattle fed a basal silage diet. J. Agric. Sci., Camb. 100:329.

Roth, F. X., and M. Kirchgessner. 1980. Contribution of dietary nucleic acids to the N metabolism. Paper 9. Proc. 3rd EAAP Symposium on Protein Metabolism and Nutrition. Braunschweig.

Roy, J. H. B., C. C. Balch, E. L. Miller, E. R. Ørskov, and R. H. Smith. 1977. Calculation of the N-requirement for ruminants from nitrogen metabolism studies. P. 126 *in* Protein Metabolism and Nutrition. Proc. 2nd Int. Symp. EAAP Pub. No. 22. Wageningen, The Netherlands: Centre for Agricultural Publishing and Documentation.

Russell, J. B., and R. E. Hespell. 1981. Microbial rumen fermentation. J. Dairy Sci. 64:1153.

Russell, J. B., and C. J. Sniffen. 1984. Effect of carbon-4 and carbon-5 volatile fatty acids on growth of mixed rumen bacteria in vitro. J. Dairy Sci. 67:987

Russell, J. B., F. J. Delfino, and R. L. Baldwin. 1979. Effects of combinations of substrates on maximum growth rates of several rumen bacteria. Appl. Environ. Microbiol. 37:544.

Russell, J. B., C. J. Sniffen, and P. J. Van Soest. 1983. Effect of carbohydrate limitation on degradation and utilization of casein by mixed rumen bacteria. J. Dairy Sci. 66:763.

Salter, D. N., and R. H. Smith. 1977. Digestibilities of nitrogen compounds in rumen bacteria and in other compounds of digesta in the small intestine of the young sheep. Br. J. Nutr. 38:207.

Santos, K. A. S. 1981. Effect of diet on amino acid flow to the ruminant intestine. Ph.D. thesis. University of Wisconsin, Madison.

Santos, K. A. S., M. D. Stern, and L. D. Satter. 1981. Ruminal protein degradation and amino acid absorption in the small intestine of lactating cows fed various protein sources. J. Anim. Sci. 55 (Suppl. 1):428.

Santos, K. A., M. D. Stern, and L. D. Satter. 1982. Protein degradation in the rumen and amino acid absorption in the small intestine of lactating dairy cattle fed various protein sources. J. Anim. Sci. 58:244.

Satter, L. D. 1982. A metabolizable protein system keyed to ruminal ammonia concentration—the Wisconsin system. P. 245 *in* Protein Requirements for Cattle: Proceedings of an International Symposium, F. N. Owens, ed. MP-109. Stillwater, Okla.: Division of Agriculture, Oklahoma State University.

Satter, L. D., and R. E. Roffler. 1975. Nitrogen requirement and utilization in dairy cattle. J. Dairy Sci. 58:1219.

Satter, L. D., and L. L. Slyter. 1974. Effect of ammonia concentration of rumen microbial protein production *in vitro*. Br. J. Nutr. 32:199.

Schaefer, D. M., C. L. Davis, and M. P. Bryant. 1980. Ammonia saturation constants for predominant species of rumen bacteria. J. Dairy Sci. 63:1248.

Schaetzel, W. P., and D. E. Johnson. 1981. Nicotinic acid and dilution rate effects on *in vitro* fermentation efficiency. J. Anim. Sci. 53:1104.

Scharrer, E. 1978. Comparative amino acid uptake at the serosal side of colon mucosa. Pflugers Arch. 376:245.

Scharrer, E., and B. Amann. 1980. Intestinal transport of amino acids and pyrimidines in sheep. Paper 12. Proc. 3rd EAAP Symp. on Protein Metabolism and Nutrition. Braunschweig.

Schneider, B. H. 1947. Feeds of the World, their Digestibility and Composition. W. Va. Agric. Exp. Stn., Morgantown.

Schoeman, E. A., P. J. DeWet, and W. J. Burger. 1972. The evaluation of the digestibility of treated proteins. Agroanimalia 4:35.

Schurch, A. 1980. The nutritional value of protein. Proc. 3rd EAAP Symp. on Protein Metabolism and Nutrition. EAAP Publ. 27:7.

Scott, H. W., and B. A. Dehority. 1965. Vitamin requirements of several cellulolytic rumen bacteria. J. Bacteriol. 89:1169.

Sharma, H. R., J. R. Ingalls, and R. J. Parker. 1974. Effects of treating rape seed meal and casein with formaldehyde on the flow of nutrients through the gastrointestinal tract of fistulated Holstein steers. Can. J. Anim. Sci. 54:305.

Siddons, R. C., and D. E. Beever. 1977. Prediction of the protein value of forage-based diets. P. 83 *in* Grassland Research Institute Annual Report.

Siddons, R. C., D. E. Beever, and R. A. Terry. 1976. Prediction of the protein value of foraged-based diets. P. 88 *in* Grassland Research Institute Annual Report.

Siddons, R. C., D. E. Beever, J. V. Nolan, A. B. McAllan, and J. C. Macrae. 1979. Estimation of microbial protein in duodenal digesta. Ann. Rech. Vet. 10:286.

Slade, L. M., D. W. Robinson, and K. E. Casey. 1970. Nitrogen metabolism in nonruminant herbivores. I. The influence of nonprotein nitrogen and protein quality on the nitrogen retention of adult mares. J. Anim. Sci. 30:753.

Slade, L. M., R. Bishop, J. G. Morris, and D. W. Robinson. 1971. Digestion and absorption of ^{15}N labelled microbial protein in the large intestine of the horse. Br. Vet. J. 127:xi.

Smith, R. H. 1975. Nitrogen metabolism in the rumen and the composition and nutritive value of nitrogen compounds entering the duodenum. P. 399 *in* Digestion and Metabolism in the Ruminant, I. W. McDonald and A. C. I. Warner, eds. Armidale, NSW, Australia: The University of New England Publishing Unit.

Smith, R. H. 1979. Synthesis of microbial nitrogen compounds in the rumen and their subsequent digestion. J. Anim. Sci. 49:1604.

Smith, R. H., and O. E. Mohamed. 1977. Effect of degradation in the rumen on dietary protein entering the ruminant duodenum. Proc. Nutr. Soc. 36:153A.

Smith, R. H., D. N. Salter, J. D. Sutton, and A. B. McAllan. 1974. Synthesis and digestion of microbial nitrogen compounds and VFA production by the bovine. P. 81 *in* Tracer Studies on Non-Protein Nitrogen for Ruminants. III. Vienna: International Atomic Energy Agency.

Sniffen, C. J. 1974. Nitrogen utilization as related to solubility of NPN and protein in feeds. P. 12 *in* Proc. Cornell Nutr. Conf.

Sniffen, C. J., and D. R. Jacobson. 1975. Net amino absorption in steers fed alfalfa hay cut at two stages of maturity. J. Dairy Sci. 58:371.

Spears, J. W., D. G. Ely, and L. P. Bush. 1978. Influence of supplemental sulfur on *in vitro* and *in vivo* microbial fermentation of Kentucky 31 tall fescue. J. Anim. Sci. 47:552.

Stallcup, O. T., G. V. Davis, and L. Shields. 1975. Influence of dry matter and nitrogen intakes on fecal nitrogen losses in cattle. J. Dairy Sci. 58:1301.

Steinhour, W. D., and J. H. Clark. 1982. Microbial nitrogen flow to the small intestine of ruminants. Pp. 166-182 *in* Protein Requirements for Cattle: Proceedings of an International Symposium, F. N. Owens, ed. MP-109. Stillwater, Okla.: Division of Agriculture, Oklahoma State University.

Stern, M. D., and W. H. Hoover. 1979. Methods for determining and factors affecting rumen microbial protein synthesis: A review. J. Anim. Sci. 49:1590.

Stern, M. D., and L. D. Satter. 1982. *In vivo* estimation of protein degradability in the rumen. P. 57 *in* Protein Requirements for Cattle: Proceedings of an International Symposium, F. N. Owens, ed. MP-109. Stillwater, Okla.: Division of Agriculture, Oklahoma State University.

Stern, M. D., and L. D. Satter. 1983. Evaluation of nitrogen solubility and the dacron bag technique as methods for estimating protein degradability in the rumen. J. Anim. Sci. 58:714.

Stern, M. D., M. E. Ortega, and L. D. Satter. 1980. Use of the dacron bag technique with rate of passage information to estimate protein degradation in the rumen. J. Anim. Sci. 51(Suppl. 1):392.

Stern, M. D., M. E. Ortega, and L. D. Satter. 1983a. Retention time in the rumen and degradation of protein supplements fed to lactating dairy cattle. J. Dairy Sci. 66:1264.

Stern, M. D., L. M. Rode, R. W. Prange, R. H. Stauffacher, and L. D. Satter. 1983b. Ruminal protein degradation of corn gluten meal in lactating dairy cattle fitted with duodenal T-type cannulae. J. Anim. Sci. 56:194.

Stern, M. D., K. A. Santos, and L. D. Satter. 1984. Protein degradation and amino acid absorption. J. Dairy Sci. (in press).

Stevens, C. E., R. A. Argenzio, and E. T. Clemens. 1980. Microbial digestion: Rumen versus large intestine. Pp. 685-706 *in* Digestive Physiology and Metabolism in Ruminants, Y. Ruckebusch and P. Thivend, eds. Westport, Conn.: AVI Publ. Co.

Stewart, C. S. 1975. Some effects of phosphate and volatile fatty acid salts on growth or rumen bacteria. J. Gen. Microbiol. 89:319.

Storm, E., and E. R. Ørskov. 1979. Utilization of rumen bacteria by ruminants. Ann. Rech. Vet. 10:294.

Storm, E., E. R. Ørskov, and R. Smart. 1983. The nutritive values of rumen microorganisms in ruminants. 2. The apparent digestibility and net utilization of microbial N for growing lambs. Br. J. Nutr. 50:471.

Stouthamer, A. H. 1973. A theoretical study on the amount of ATP required for synthesis of microbial cell material. Antonie van Leeuwenhoek. J. Microbiol. Seral. 39:945.

Stouthamer, A. H. 1979. The search for correlation between theoretical and experimental growth yields. Int. Rev. Biochem. Microb. Biochem. 21:1.

Stouthamer, A. H., and C. Bettenhaussen. 1975. Utilization of energy for growth and maintenance in continuous and batch cultures of microorganisms. Biochim. Biophys. Acta 301:53.

Strozinski, L. L., and P. T. Chandler. 1972. Nitrogen metabolism and metabolic fecal nitrogen as related to caloric intake and digestibility. J. Dairy Sci. 55:1281.

Sutton, J. D., R. H. Smith, A. B. McAllan, J. E. Storry, and D. A. Corse. 1975. Effect of variations in dietary protein and of supplements of cod-liver oil on energy digestion and microbial synthesis in the rumen of sheep fed hay and concentrates. J. Agric. Sci., Camb. 84:317.

Swanson, E. W. 1977. Factors for computing requirements of protein for maintenance of cattle. J. Dairy Sci. 60:1583.

Swanson, E. W. 1982. Estimation of metabolic protein requirements to cover unavailable losses of endogenous nitrogen in maintenance of cattle. P. 183 *in* Protein Requirements of Cattle: Proceedings of an International Symposium, F. N. Owens, ed. MP-109. Stillwater, Okla.: Division of Agriculture, Oklahoma State University.

Tagari, H. 1969. Comparison of the efficiency of proteins contained in lucerne hay and soya-bean meal for sheep. Br. J. Nutr. 23:455.

Tagari, H., and E. N. Bergman. 1978. Intestinal disappearances and portal blood appearance of amino acids in sheep. J. Nutr. 108:790.

Tagari, H., I. Ascarelli, and A. Bondi. 1962. The influence of heating on the nutritive value of soya-bean meal for ruminants. Br. J. Nutr. 16:237.

Tamminga, S. 1979. Protein degradation in the forestomachs of ruminants. J. Anim. Sci. 49:1615.

Tamminga, S. 1980. Amino acid supply and utilization in ruminants. Paper 42. Proc. 3rd EAAP Symp. on Protein Metabolism and Nutrition. Braunschweig.

Tamminga, S. 1981a. Effect of the roughage/concentrate ratio on nitrogen entering the small intestine of dairy cows. Neth. J. Agric. Sci. 29:273.).

Tamminga, S. 1981b. Effect of frequency of feeding concentrate diets on N entering the small intestine of dairy cows. P. 87 *in* Proc. Amino Acid Symposium, Warsaw.

Tamminga, S., and J. D. Oldham. 1980. Amino acid utilization by dairy cows. II. Concept of amino acid requirements. Livestock Prod. Sci. 7:453.

Tamminga, S., and K. K. van Hellemond. 1977. The protein requirements of dairy cattle and developments in the use of protein, essential amino acids and non-protein nitrogen, in the feeding of dairy cattle. P. 9 *in* Protein and Non-Protein Nitrogen for Ruminants. Recent Developments in the Use of New Sources. Elmsford, N.Y.: Pergamon Press.

Tamminga, S., C. J. Van Der Koelen, and A. M. Van Vuuren. 1979. Effect of level of feed intake on nitrogen entering the small intestine of dairy cows. Livestock Prod. Sci. 6:255.

Tas, M. V., R. A. Evans, and R. F. E. Exford. 1981. The digestibility of amino acids in the small intestine of the sheep. Br. J. Nutr. 45:167.

Taylor, R. B. 1962. Pancreatic secretion in the sheep. Res. Vet. Sci. 3:63.

Theurer, C. B. 1979. Microbial protein synthesis as influenced by diet. *In* Regulation of Acid-Base Balance. Piscataway, N.J.: Church and Dwight Co.

Theurer, B., W. Woods, and G. E. Poley. 1966. Comparison of portal and jugular blood plasma amino acids in lambs at various intervals postprandial. J. Anim. Sci. 25:175.

Thomas, J. W., Y. Yu, T. Middleton, and C. Stallings. 1982. Estimations of protein damage. P. 81 *in* Protein Requirements of Cattle: Proceedings of an International Symposium, F. N. Owens, ed. MP-109. Stillwater, Okla.: Division of Agriculture, Oklahoma State University.

Thomas, P. C. 1973. Microbial protein synthesis. Proc. Nutr. Soc. 32:85.

Thomson, D. J., D. E. Beever, J. J. Lathan, M. E. Sharpe, and R. A. Terry. 1977. The effect of inclusion of mineral salts in the diet on dilution rate, the pattern of rumen fermentation and the composition of rumen microflora. J. Agric. Sci., Camb. 91:1.

Thomson, D. J., D. E. Beever, C. R. Lonsdale, M. J. Haines, S. B. Cammell, and A. R. Austin. 1981. The digestion by cattle of grass silage made with formic acid and formic acid-formaldehyde. Br. J. Nutr. 46:193.

Thornton, R. F., P. R. Bird, M. Sommers, and R. J. Moir. 1970. Urea excretion in ruminants. III. The role of the hind-gut (caecum and colon). Aust. J. Agric. Res. 21:345.

Trenkle, A. 1982. The metabolizable protein feeding standard. P. 238 *in* Protein Requirements for Cattle: Proceedings of an International Symposium, F. N. Owens, ed. MP-109. Stillwater, Okla.: Division of Agriculture, Oklahoma State University.

Tyrrell, H. F., and P. W. Moe. 1975. Effect of intake on digestive efficiency. J. Dairy Sci. 58:1151.

Tyrrell, H. F., P. W. Moe, and R. R. Oltjen. 1974. Energetics of growth and fattening compared to lactation in cattle. *In* Energy Metabolism of Farm Animals, K. H. Menke, H. J. Lanzsch, and J. R. Reichl, eds. European Assoc. Anim. Prod. Publ. No. 14. Proc. 6th Symp. Energy Metabolism, Stuttgart.

Ulyatt, M. J., D. W. Dellow, C. S. W. Reid, and T. Bauchop. 1975a. Structure and function of the large intestine of ruminants. Pp. 119–133 *in* Digestion and Metabolism in the Ruminant, I. W. McDonald and A. C. I. Warner, eds. Armidale, NSW, Australia: The University of New England Publishing Unit.

Ulyatt, M. J., J. C. MacRae, R. T. J. Clarke, and P. D. Pearce. 1975b. Quantitative digestion of fresh herbage by sheep. IV. Protein synthesis in the stomach. J. Agric. Sci., Camb. 84:453.

Van der Aar, P. J., L. L. Berger, and G. C. Fahey, Jr. 1982. The effect of alcohol treatments on solubility and *in vitro* and *in situ* digestibilities of soybean meal protein. J. Anim. Sci. 55:1179.

Van't Klooster, A. Th., and H. A. Boekholt. 1972. Protein digestion in the stomachs and intestines of the cow. Neth. J. Agric. Sci. 20:272.

Van Nevel, C. J., and D. I. Demeyer. 1976. Determination of rumen microbial growth *in vitro* from ^{32}P-labelled phosphate incorporation. Br. J. Nutr. 38:101.

Van Soest, P. J. 1982. Nutritional Ecology of the Ruminant. Corvallis, Oreg.: O & B Books, Inc.

Van Soest, P. J., J. Fadel, and C. J. Sniffen. 1979. Discount factors for energy and protein in ruminant feeds. P. 63 *in* Proc. Cornell Nutr. Conf.

Van Soest, P. J., C. J. Sniffen, D. R. Mertens, D. G. Fox, P. H. Robinson, and U. Krishnamoorthy. 1982. A net protein system for cattle: The rumen submodel for nitrogen. P. 265 *in* Protein Requirements for Cattle: Proceedings of an International Symposium, F. N. Owens, ed. MP-109. Stillwater, Okla.: Division of Agriculture, Oklahoma State University.

Varga, G. A., and E. C. Prigge. 1982. Influence of forage species and level of intake on ruminal turnover rates. J. Anim. Sci. 55:1498.

Veira, D. M., G. K. MacLeod, J. H. Burton, and J. B. Stone. 1980. Nutrition of the weaned Holstein calf. I. Effect of dietary protein level on rumen metabolism. J. Anim. Sci. 50:937.

Vérité, R. 1980. Appreciation of the nitrogen value of feeds for ruminants. Page 87 *in* Standardization of Analytical Methodology for Feeds. Ottawa, Ont.: Int. Dev. Res. Ctr.

Vérité, R., and C. Demarquilly. 1978. Qualite des matieres azotees des aliments pour ruminants. P. 143 *in* La Vache Laitiere. Versailles: INRA Publications.

Vérité, R., and D. Sauvant. 1981. Prevision de la valeur nutritive azotee des aliments concentres pour la ruminants. P. 279 *in* Prevision de la nutritive des aliments des ruminants. Versailles: INRA Publications.

Vérité, R., M. Journet, L. Gueguen, and A. Hoden. 1978. Vaches laitieres. P. 345 *in* Alimentation des Ruminants. Versailles: INRA Publications.

Vérité, R., M. Journet, and R. Jarrige. 1979. A new system for the protein feeding of ruminants: The PDI system. Livestock Prod. Sci. 6:349.

Vogels, G. D., W. F. Hoppe, and C. K. Stumm. 1980. Associaton of methanogenic bacteria with rumen ciliates. Appl. Environ. Microbiol. 40:608.

Wagner, D. G., and J. K. Loosli. 1967. Studies on the energy requirements of high-producing dairy cows. Memoir No. 400. Cornell Univ. Agric. Exp. Stn.

Waldo, D. R. 1977a. Potential of chemical preservation and improvement of forages. J. Dairy Sci. 60:306.

Waldo, D. R. 1977b. Silage and supplement effects on intake and growth by dairy heifers. P. 266 *in* Abstracts 59th Annual Meeting. Am. Soc. Anim. Sci.

Waldo, D. R. 1979. Meeting the ruminant's protein need in the future. P. 27 *in* Proc. 33rd Annual Virginia Feed Convention and Nutr. Conf.

Waldo, D. R., and H. K. Goering. 1979. Insolubility of proteins in ruminant feeds by four methods. J. Anim. Sci. 49:1560.

Waldo, D. R., and B. P. Glenn. 1982. Foreign systems for meeting the protein requirements of ruminants. P. 296 *in* Protein Requirements of Cattle: Proceedings of an International Symposium, F. N. Owens, ed. MP-109. Stillwater, Okla.: Division of Agriculture, Oklahoma State University.

Waldo, D. R., and N. A. Jorgensen. 1981. Forages for high animal production: Nutritional factors and effects of conservation. J. Dairy Sci. 64:1207.

Waldo, D. R., and H. F. Tyrrell. 1980. The relation of insoluble nitrogen intake to gain, energy retention and nitrogen retention in Holstein steers. P. 572 *in* Proc. 3rd EAAP Symp. on Protein Metabolism and Nutrition. Braunschweig.

Walker, D. J., A. R. Egan, C. J. Nader, M. J. Ulyatt, and G. B. Storer. 1975. Rumen microbial protein synthesis and proportions of microbial and nonmicrobial nitrogen flowing to the intestines of sheep. Aust. J. Agric. Res. 26:699.

Waller, J. C., J. R. Black, and W. G. Bergen. 1982. Michigan protein system(s). P. 323 *in* Protein Requirements for Cattle: Proceedings of an International Symposium, F. N. Owens, ed. MP-109. Stillwater, Okla.: Division of Agriculture, Oklahoma State University.

Warner, A. C. I. 1981. Rate of passage of digesta through the gut of mammals and birds. Nutr. Abs. Rev. 51:789 (Series B).

Weakley, D. C., F. N. Owens, K. B. Poling, and C. E. Kautz. 1983. Influence of roughage level on soybean meal degradation and microbial protein synthesis in the rumen. P. 1 *in* Anim. Sci. Res. Report. Stillwater, Okla.: Oklahoma State University.

Whetstone, H. D., C. L. Davis, and M. P. Bryant. 1981. Effect of monensin on breakdown of protein by ruminal microorganisms *in vitro*. J. Anim. Sci. 53:803.

Whitelaw, F. G., and T. R. Preston. 1963. The nutrition of the early weaned calf. III. Protein solubility and amino acid composition as factors affecting protein utilization. Anim. Prod. 5:131.

Whitelaw, F. G., J. M. Eadie, S. O. Mann, and R. S. Reid. 1972. Some effects of rumenciliate protozoa in cattle given restricted amounts of a barley diet. Br. J. Nutr. 27:425.

Whitlow, L. W. 1979. Rumen microbial degradation of feed protein. Ph.D. thesis. University of Wisconsin.

Whitlow, L. W., and L. D. Satter. 1979. Evaluation of models which predict amino acid flow to the intestine. Ann. Rech. Vet. 10:307.

Williams, V. J. 1969. The relative rates of absorption of amino acids from the small intestine of the sheep. Comp. Biochem. Physiol. 29:865.

Wohlt, J. E., C. J. Sniffen, and W. H. Hoover. 1973. Measurement of protein solubility in common feedstuffs. J. Dairy Sci. 56:1052.

Wohlt, J. E., C. J. Sniffen, W. H. Hoover, and L. L. Johnson. 1976. Nitrogen metabolism in wethers as affected by dietary protein solubility and amino acid profile. J. Anim. Sci. 42:1280.

Wolff, J. E., E. N. Bergman, and H. H. Williams. 1972. Net metabolism of plasma amino acids by liver and portal-drained viscera of fed sheep. Am. J. Physiol. 223:438.

Wolfrom, G. W., J. M. Asplund, and T. R. Hoover. 1979. Effect of portal versus jugular methionine infusion on circulating amino acids and nitrogen metabolism in sheep. J. Nutr. 109:1979.

Wolstrup, J., K. Jensen, and A. Just. 1979. ATP and DNA as microbial parameters in the alimentary tract. Ann. Rech. Vet. 10:283.

Wrong, O. M., C. J. Edwards, and V. S. Chadwick. 1981. The Large Intestine: Its Role in Mammalian Nutrition and Homeostasis. New York: John Wiley & Sons.

Zinn, R. A., and F. N. Owens. 1980. Influence of roughage level and feed intake level on digestive function. Okla. Res. Update 107:150.

Zinn, R. A., and F. N. Owens. 1981a. Factors influencing protein digestion in ruminants. J. Anim. Sci. 51(Suppl. 10):412.

Zinn, R. A., and F. N. Owens. 1981b. Influence of level of feed intake on nitrogen metabolism in steers fed high concentrate rations. P. 448 *in* Am. Soc. Anim. Sci. (Abstr.)

Zinn, R. A., and F. N. Owens. 1982. Predicting net uptake of nonammonia N from the small intestine. P. 133 *in* Protein Requirements of Cattle: Proceedings of an International Symposium, F. N. Owens, ed. MP-109. Stillwater, Okla.: Division of Agriculture, Oklahoma State University.

Zinn, R. A., and F. N. Owens. 1983. Site of protein digestion in steers: predictability. J. Anim. Sci. 56:707.

Zinn, R. A., L. S. Bull, R. W. Hemken, F. S. Button, C. Enlow, and R. W. Tucker. 1980. Apparatus for measuring and subsampling digesta in ruminants equipped with reentrant intestinal cannulas. J. Anim. Sci. 51:193.

Zinn, R. A., L. S. Bull, and R. W. Hemken. 1981. Degradation of supplemental proteins in the rumen. J. Anim. Sci. 52:857.

Appendix Tables

APPENDIX TABLE 1 Numerical Data from the Protein Systems Used in Figures 3 Through 12

System[a]	Milk	Protein	Undegrad-ability	Predicted[b]	Tamminga and van Hellemond (1977)	Rohr et al. (1979)	Journet and Verite (1979)	Feces	Urine	Milk
	kg	% DM	%			(kg/day)			(% dietary)	
A	10	9.28	7.2	0.714	1.294	1.057	1.087	27.2	38.6	34.2
A	20	10.85	20.7	1.286	1.954	1.700	1.726	27.6	33.1	39.3
A	30	11.59	25.8	1.857	2.628	2.356	2.371	27.8	31.0	41.3
A	40	12.00	28.3	2.429	3.315	3.024	3.022	27.8	29.8	42.3
B	10	9.39	7.3	0.835	1.422	1.182	1.188	31.1	37.6	31.2
B	20	10.21	14.7	1.308	2.035	1.779	1.736	29.4	30.0	40.6
B	30	10.66	18.3	1.781	2.649	2.376	2.283	28.6	26.3	45.1
B	40	10.94	20.4	2.255	3.263	2.973	2.831	28.1	24.1	47.7
C	10	11.12	41.3	1.071	1.422	1.182	1.444	31.1	41.7	27.2
C	20	12.87	49.2	1.786	2.035	1.779	2.184	30.2	36.6	33.2
C	30	13.82	52.7	2.501	2.649	2.376	2.923	29.9	34.3	35.8
C	40	14.42	54.7	3.216	3.263	2.973	3.663	29.7	33.0	37.3
D	10	13.23	13.9	1.033	1.270	1.034	1.139	28.4	45.3	26.3
D	20	15.49	26.5	1.845	1.893	1.640	1.846	26.2	43.3	30.5
D	30	16.66	31.6	2.657	2.516	2.246	2.552	25.3	42.5	32.2
D	40	17.37	34.4	3.469	3.139	2.852	3.259	24.7	42.1	33.2
K	10	10.62	26.2	0.996	1.236	1.001	1.189	41.5	28.5	30.0
K	20	12.81	33.7	1.667	1.853	1.601	1.850	36.6	27.0	36.4
K	30	14.08	36.9	2.338	2.469	2.201	2.510	34.5	26.3	39.2
K	40	16.05	37.7	2.951	3.086	2.801	3.135	32.1	26.4	41.5
L	10	9.00	17.1	0.976	1.397	1.158	1.245	48.0	26.4	25.6
L	20	11.92	25.2	1.622	2.078	1.820	1.916	41.8	26.7	31.5
L	30	13.91	28.9	2.267	2.758	2.482	2.588	39.1	26.8	34.0
L	40	15.36	30.9	2.913	3.439	3.145	3.259	37.6	26.9	35.5
N	10	11.99	--	1.307	1.422	1.182	--	43.7	31.0	25.2
N	20	13.77	--	2.127	2.035	1.779	--	41.3	27.7	31.0
N	30	14.75	--	2.947	2.649	2.376	--	40.2	26.2	33.6
N	40	15.37	--	3.767	3.263	2.973	--	39.6	25.4	35.0
P	10	10.26	34.7	1.223	1.318	1.082	1.360	47.7	28.5	23.8
P	20	13.15	37.7	1.899	1.968	1.713	2.037	37.8	31.3	30.9
P	30	15.35	37.7	2.580	2.691	2.417	2.747	32.2	33.7	34.1
P	40	17.01	37.7	3.261	3.414	3.120	3.458	29.0	35.0	36.0
S	10	10.25	17.8	0.971	1.422	1.182	1.263	18.7	54.5	26.8
S	20	11.52	26.9	1.572	2.035	1.779	1.884	18.1	48.2	33.7
S	30	12.22	31.1	2.172	2.649	2.376	2.505	17.8	45.3	36.9
S	40	12.66	33.4	2.773	3.263	2.973	3.126	17.6	43.7	38.7

[a]A, ARC (1980); B, Burroughs (1971; 1974; 1975a,b); C, Chalupa (1975a, 1980a); D, Danfaer (1979); K, Kaufmann (1977b, 1979); L, Landis (1979); N, NRC (1978); P, PDI (Verite et al., 1979) and S, Satter (1982).

[b]Predicted protein flows into the small intestine were calculated as the sum of true microbial and protozoal protein plus undegraded intake protein without any endogenous protein for each system.

[c]Expected protein flows into the small intestine were calculated for each system based on equations using either digestible organic matter intake (Tamminga and van Hellemond, 1977; Rohr et al., 1979) or digestible organic matter intake and undegraded intake protein (Journet and Verite, 1979).

APPENDIX TABLE 2   *In Vivo* Estimates of the Percentage of Undegraded Protein in Common Feedstuffs

Feedstuff	Animal	Basic Diet	Level of Intake, % of BW	Fraction of Undegraded Protein	Reference
Cereals					
Barley	Sheep	All Barley (rolled)	2.8	0.14	Mathers and Miller, 1981
Barley	Sheep	All Barley	∼1.2	0.28	Mathison and Milligan, 1971
Corn	Angus Steers	74% Dry Rolled Corn 20% Chopped Prairie Hay	1.9	0.58	Zinn and Owens, 1983
Corn	Angus Steers (203 kg)	74% Dry Rolled Corn 20% Chopped Prairie Hay	1.9	0.73	Zinn and Owens, 1983
Corn	Angus Steers (530 kg)	80% Corn Grain (15–35% moisture) 14% Cottonseed Hulls 6% Pelleted Supplement	1.8	0.64	Aguirre et al., 1984
Sorghum Grain (Dry Ground)	Angus Steers	83% Sorghum Grain 15% Coastal Bermuda Grass	∼1.4	0.49	Potter et al., 1971
Sorghum Grain (Reconstituted)	Angus Steers	83% Sorghum Grain 15% Coastal Bermuda Grass	∼1.4	0.20	Potter et al., 1971
Sorghum Grain (Steam Flaked)	Angus Steers	83% Sorghum Grain 15% Coastal Bermuda Grass	∼1.4	0.42	Potter et al., 1971
Sorghum Grain (Micronized)	Angus Steers	83% Sorghum Grain 15% Coastal Bermuda Grass	∼1.4	0.64	Potter et al., 1971
Sorghum Grain (Dry Rolled)	Beef Steers	82% Sorghum Grain	?	0.58	Theurer, 1979
Sorghum Grain (Steam Processed Flaked)	Beef Steers	82% Sorghum Grain	?	0.52	Theurer, 1979
Sorghum Grain (Dry Rolled)	Angus & Hereford Steers (350 kg)	88% Sorghum Grain 8% Cottonseed Hulls	2.0	0.69	Hibberd, 1982
Sorghum Grain (Reconstituted)	Angus & Hereford Steers (350 kg)	88% Sorghum Grain 8% Cottonseed Hulls	2.0	0.65	Hibberd, 1982
Zein	Sheep	Partially Purified	∼1.7	0.60	McDonald, 1954
Oil Meals					
Cottonseed Meal (Solvent)	Holstein Steers (179 kg)	40% Chopped Alfalfa Hay 60% Concentrate	1.7	0.24	Zinn et al., 1981

APPENDIX TABLE 2   *(Continued)*

Feedstuff	Animal	Basic Diet	Level of Intake, % of BW	Fraction of Undegraded Protein	Reference
Oil Meals (cont.)					
Cottonseed Meal (Solvent)	Holstein Steers (179 kg)	40% Chopped Alfalfa Hay 60% Concentrate	2.2	0.61	Zinn et al., 1981
Cottonseed Meal (Solvent)	Angus Steers (203 kg)	74% Dry Rolled Corn 20% Chopped Prairie Hay	1.9	0.50	Zinn and Owens, 1983
Cottonseed Meal (Solvent)	Angus Steers (203 kg)	60% Chopped Prairie Hay 16% Dry Rolled Corn 10% Soybean Meal	1.9	0.43	Zinn and Owens, 1983
Cottonseed Meal (Solvent)	Holstein & Ayrshire Cows (463 kg)	42% Ground Corn 20% Alfalfa Hay 20% Prairie Hay 15% Cottonseed Meal	3.1	0.35	Goetsch and Owens, 1985
Cottonseed Meal (Solvent)	Crossbred Steers (322 kg)	59% Dry Rolled Corn 20% Prairie Hay 15% Cottonseed Meal	1.7	0.34	Goetsch and Owens, 1985
Cottonseed Meal (Prepress)	Holstein & Ayrshire Cows (463 kg)	42% Ground Corn 20% Alfalfa Hay 20% Prairie Hay 15% Cottonseed Meal	3.1	0.35	Goetsch and Owens, 1985
Cottonseed Meal (Prepress)	Crossbred Steers (322 kg)	59% Dry Rolled Corn 20% Prairie Hay 15% Cottonseed Meal	1.7	0.38	Goetsch and Owens, 1985
Cottonseed Meal (Screw Press)	Holstein & Ayrshire Cows (463 kg)	42% Ground Corn 20% Alfalfa Hay 20% Prairie Hay 15% Cottonseed Meal	3.1	0.57	Goetsch and Owens, 1985
Cottonseed Meal (Screw Press)	Crossbred Steers (322 kg)	59% Dry Rolled Corn 20% Prairie Hay 15% Cottonseed Meal	1.7	0.43	Goetsch and Owens, 1985
Linseed Meal	Holstein Steers (179 kg)	40% Chopped Alfalfa Hay 60% Concentrate	2.2	0.44	Zinn et al., 1981
Peanut Meal	Sheep	Barley	Maintenance	0.22	Miller, 1973
Peanut Meal	Merino Wethers	40% Oat Hulls 25% Peanut Meal 17% Starch 9% Sucrose	~1.6	0.37	Hume, 1974

APPENDIX TABLE 2   *(Continued)*

Feedstuff	Animal	Basic Diet	Level of Intake, % of BW	Fraction of Undegraded Protein	Reference
Oil Meals (cont.)					
Rapeseed Meal	Jersey Heifers (250 kg)	80% Grass Silage 20% Soybean Meal	1.7	0.23	Rook et al., 1983
Soybean Meal	Holstein Steers (179 kg)	40% Chopped Alfalfa Hay 60% Concentrate	1.7	0.15	Zinn et el., 1981
Soybean Meal	Holstein Steers (179 kg)	40% Chopped Alfalfa Hay 60% Concentrate	2.2	0.18	Zinn et el., 1981
Soybean Meal	Merino Wethers	40% Oat Hulls 21% Starch 21% Soybean Meal 12% Sucrose	~1.6	0.61	Hume, 1974
Soybean Meal	Angus Steers (475 kg)	53% Corn Cobs 17% Cornstarch 16% Brewers Grains	1.8	0.24	Merchen et al., 1979
Soybean Meal	Steers (350 kg)	75% Cottonseed Hulls 20% Milo	1.4	0.20	Kropp et al., 1977a
Soybean Meal	Angus Steers (203 kg)	74% Dry Rolled Corn 20% Chopped Prairie Hay	1.9	0.43	Zinn and Owens, 1983
Soybean Meal	Angus Steers (203 kg)	60% Chopped Prairie Hay 16% Dry Rolled Corn 10% Soybean Meal	1.9	0.24	Zinn and Owens, 1983
Soybean Meal	Jersey Heifers (250 kg)	86% Grass Silage 14% Soybean Meal	1.6	0.10	Rook et al., 1983
Soybean Meal	Angus/ Hereford Steers (320 kg)	46% Corn Cobs 22% Cornstarch Grits 10% Ensiled Cornstalks 12% Soybean Meal	1.6	0.29	Loerch et al, 1983
Soybean Meal	Holstein & Ayrshire Cows (463 kg)	42% Ground Corn 20% Alfalfa Hay 20% Prairie Hay 12% Soybean Meal	3.1	0.35	Goetsch and Owens, 1985
Sunflower Meal	Lambs	Barley?	1.25 X Maintenance	0.28	Miller, 1973
Sunflower Meal	Lambs	Barley?	2.5 X Maintenance	0.19	Miller, 1973

APPENDIX TABLE 2   (*Continued*)

Feedstuff	Animal	Basic Diet	Level of Intake, % of BW	Fraction of Undegraded Protein	Reference
By Product Feeds					
Blood Meal	Angus/ Hereford (320 kg)	54% Corn Cobs 28% Cornstarch Grits 10% Ensiled Cornstalks	1.6	0.82	Loerch et al., 1983
Brewers Dried Grains	Angus Steers (475 kg)	53% Corn Cobs 17% Cornstarch 16% Brewers Grains	1.8	0.48	Merchen et al., 1979
Brewers Dried Grains	Angus Steers (262 kg)	45% Corn Cobs 31% Sorghum Grain 16% Brewers Grains	2.4	0.61	Merchen et al., 1979
Brewers Dried Grains	Holstein Heifers (450 kg)	Oat Straw Ground Oats Dehydrated Alfalfa Cornstarch	1.8	0.27	Santos et al., 1981
Brewers Dried Grains	Steers (262 kg)	Corn Cobs Sorghum Grain Brewers Dried Grains	2.7	0.61	Poos et al., 1979b
Brewers Dried Grains	Wethers (49 kg)	34% Ground Oats 15% Dehydrated Alfalfa 0-25 Oat Straw 0-20 Cornstarch	2.1	0.66	Whitlow, 1979
Corn Gluten Meal	Holstein Steers (179 kg)	40% Chopped Alfalfa Hay 60% Concentrate	1.7	0.46	Zinn et al., 1981
Corn Gluten Meal	Holstein Steers (179 kg)	40% Chopped Alfalfa Hay 60% Concentrate	2.2	0.61	Zinn et al., 1981
Corn Gluten Meal	Holstein Cows	Silage Corn Grain Alfalfa Hay	3.1	0.57	Stern et al., 1983b
Distillers Dried Grains with Solubles	Holstein Heifers (400 kg)	Oat Straw Ground Oats Dehydrated Alfalfa	1.0	0.55	Santos et al., 1981
Distillers Dried Grains with Solubles	Sheep	34% Ground Oats 16% Dehydrated Alfalfa 0-45% Ground Oat Straw 0-45% Distillers Dried Grains with Solubles	2.1	0.68	Whitlow, 1979
Fish Meal	Friesian Calves (115 kg)	Corn Silage Barley-Cornstarch Fish Meal	3.0	0.78	Cottrill et al., 1982

APPENDIX TABLE 2   (*Continued*)

Feedstuff	Animal	Basic Diet	Level of Intake, % of BW	Fraction of Undegraded Protein	Reference
By Product Feeds (cont.)					
Fish Meal (Peruvian)	Sheep	Barley	Maintenance	0.69	Miller, 1973
Fish Meal (Peruvian)	Lactating Cows	Barley Barley Straw	Ad Libitum	1.00	Miller, 1973
Fish Meal	Merino Wethers	40% Oat Hulls 21% Starch 19% Fish Meal 12% Sucrose	∼1.6	0.71	Hume, 1974
Meat Meal	Angus Steers (203 kg)	74% Dry Rolled Corn 20% Chopped Prairie Hay	1.9	0.76	Zinn and Owens, 1983
Meat and Bone Meal	Holstein Steers (179 kg)	40% Chopped Alfalfa Hay 60% Concentrate	2.2	0.70	Zinn et al., 1981
Meat and Bone Meal	Angus/ Hereford Steers (320 kg)	44% Corn Cobs 24% Cornstarch Grits 10% Ensiled Cornstalks 12% Meat and Bone Meal	1.6	0.49	Loerch et al., 1983
Forages (Dry)					
Alfalfa	Sheep	All Alfalfa (Ground)	2.8	0.28	Mathers and Miller, 1981
Alfalfa	Sheep (32–42 kg)	All Alfalfa (Chopped)	∼2.9	0.20–0.24	Kennedy et al., 1982
Alfalfa	Sheep	Alfalfa	1.5	0.41	Nolan and Leng, 1972
Alfalfa	Merino Ewe	Alfalfa	∼2.0	0.21	Pilgrim et al., 1970
Alfalfa (Dehydrated)	Angus Steers (203 kg)	74% Dry Rolled Corn 20% Chopped Prairie Hay	1.9	0.57	Zinn and Owens, 1983
Alfalfa (Dehydrated)	Angus Steers (203 kg)	74% Dry Rolled Corn 20% Chopped Prairie Hay	1.9	0.62	Zinn and Owens, 1983
Alfalfa (Dehydrated)	Angus/ Hereford Steers (320 kg)	39% Dehydrated Alfalfa 24% Cornstarch Grits 16% Corn Cobs 10% Ensiled Cornstalks	1.6	0.66	Loerch et al., 1983
Alfalfa- Bromegrass	Sheep	Alfalfa- Bromegrass	∼1.3	0.21	Mathison and Milligan, 1971

APPENDIX TABLE 2 *(Continued)*

Feedstuff	Animal	Basic Diet	Level of Intake, % of BW	Fraction of Undegraded Protein	Reference
Forages (Dry) (cont.)					
Bromegrass	Sheep (32–42 kg)	Bromegrass (Chopped)	∼3.0	0.40–0.49	Kennedy et al., 1982
Lupin Meal	Merino Wethers	40% Oat Hulls 29% Lupin Meal 14% Starch 8% Sucrose	∼1.6	0.35	Hume, 1974
Ryegrass (Artificially dried)	Sheep (45–55 kg)	Ryegrass (Chopped) or (Pelleted)	∼2.0 2.0	0.30 0.54	Beever et al., 1981
Subterranean Clover (Immature)	Merino Wethers	Subterranean Clover	∼2.0	0.27	Hume and Purser, 1974
Subterranean Clover (Mature)	Merino Wethers	Subterranean Clover	∼1.6	0.52	Hume and Purser, 1974
Timothy (Artifically Dried)	Sheep (45–55 kg)	Timothy (Chopped) or (Pelleted)	∼2.0 ∼2.0	0.32 0.53	Beever and Thomson, 1981
Silages					
Corn Silage	Friesian Calves (115 kg)	Corn Silage Barley- Cornstarch Fish Meal	∼3.0	0.27	Cottrill et al., 1982
Mixed Diets					
Alfalfa (Chopped) (66%) Barley (Rolled) (33%)	Sheep	Alfalfa and Barley	2.8	0.24	Mathers and Miller, 1981
Barley (Rolled) (33%) Alfalfa (Chopped) (66%)	Sheep	Alfalfa and Barley	2.8	0.14	Mathers and Miller, 1981
Soybean Meal (50% of CP) Corn Silage Corn Grain Alfalfa Hay	Holstein Cows	Corn Silage Corn Grain Soybean Meal Alfalfa Hay	∼2.8	0.27	Stern and Satter, 1982
Soybeans (Unheated) (50% of CP) Corn Silage Corn Grain Alfalfa Hay	Holstein Cows	Corn Silage Corn Grain Soybeans Alfalfa Hay	∼2.8	0.20	Stern and Satter, 1982
Soybeans (Extruded 270°F) (50% of CP) Corn Silage Corn Grain Alfalfa Hay	Holstein Cows	Corn Silage Corn Grain Soybeans (Extruded) Alfalfa Hay	∼2.8	0.34	Stern and Satter, 1982

APPENDIX TABLE 2 *(Continued)*

Feedstuff	Animal	Basic Diet	Level of Intake, % of BW	Fraction of Undegraded Protein	Reference
Mixed Diets (cont.)					
Soybeans (Extruded 300°F) (50% of CP) Corn Silage Corn Grain Alfalfa Hay	Holstein Cows	Corn Silage Corn Grain Soybeans (Extruded) Alfalfa Hay	~2.8	0.40	Stern and Satter, 1982
Soybean Meal (50% of CP) Corn Grain Corn Silage Alfalfa Hay	Holstein Cows	Corn Grain Corn Silage Soybean Meal Alfalfa Hay	~2.5	0.30	Santos et al., 1982
Corn Gluten Meal (50% of CP) Corn Grain Corn Silage Alfalfa Hay	Holstein Cows	Corn Grain Corn Silage Corn Gluten Meal Alfalfa Hay	~2.5	0.55	Santos et al., 1982
Brewers Grains (Wet) (50% of CP) Corn Grain Corn Silage Alfalfa Hay	Holstein Cows	Corn Grain Corn Silage Brewers Grains Alfalfa Hay	~2.5	0.48	Santos et al., 1982
Distillers Dried Grains with Solubles (50% of CP) Corn Grain Corn Silage Alfalfa Hay	Holstein Cows	Corn Grain Corn Silage Distillers Dried Grains with Solubles Alfalfa Hay	~2.5	0.54	Santos et al., 1982
Alfalfa Silage (29% DM) (70% of CP) Corn Grain	Holstein Cows	Alfalfa Silage (65) Corn Grain (34)	~2.8	0.15	Merchen and Satter, 1983b
Alfalfa Silage (40% DM) (70% of CP) Corn Grain	Holstein Cows	Alfalfa Silage (65) Corn Grain (34)	~3.2	0.15	Merchen and Satter, 1983b
Alfalfa Silage (66% DM) (70% of CP) Corn Grain	Holstein Cows	Alfalfa Silage (65) Corn Grain (34)	~3.0	0.36	Merchen and Satter, 1983b
Alfalfa Silage (70% of CP) Corn Grain	Holstein Cows	Alfalfa Silage (65) Corn Grain (34)	~2.9	0.22	Merchen and Satter, 1983b
Cottonseed Meal (Solvent) (50% of CP) Corn Silage Corn Grain	Holstein Cows	Corn Silage Cottonseed Meal Corn Grain	~2.7	0.23	Pena and Satter, 1983

APPENDIX TABLE 2   (*Continued*)

Feedstuff	Animal	Basic Diet	Level of Intake, % of BW	Fraction of Undegraded Protein	Reference
Mixed Diets (cont.)					
Cottonseed Meal (Expeller) (50% of CP) Corn Silage Corn Grain	Holstein Cows	Corn Silage Cottonseed Meal Corn Grain	∼2.7	0.32	Pena and Satter, 1983
Barley (45%) Barley Straw (45%) Tapioca (10%)	Steers (150 kg)	Barley Barley Straw Tapioca	2.3 3.3	0.29 0.34	McAllan and Smith, 1983
Barley (44%) Barley Straw (44%) Soybean Meal (12%)	Steers (150 kg)	Barley Barley Straw Soybean Meal	2.3 3.3	0.28 0.31	McAllan and Smith, 1983
Barley (44%) Barley Straw (44%) Soybean Meal (12%)	Steers (150 kg)	Barley Barley Straw Soybean Flour	2.3 3.3	0.33 0.30	McAllan and Smith, 1983
Barley (ä92%) Canola Seed Meal (3%)	Sheep (32–42 kg)	Barley Canola Seed Meal	∼1.9	0.10	Kennedy et al., 1982
Cottonseed (62% of CP) Corn Silage Corn Grain	Holstein Cows	Corn Silage Cottonseed Corn Grain	∼2.2	0.44	Pena et al., 1983

APPENDIX TABLE 3   Microbial Flow and Rumen Organic Matter Digestion

OBS	REFNO		SPECIES[a]	BACTYLD gN/d	FORAGE	TDN	DE	EE	LIGNIN	DMI	OMI	RCOM	TCOM	FEEDFRQ	BODYWT	METHOD[b]
					-- % in Diet--		Mcal diet	-% in diet-		----	kg/d	-------	% DMI		kg	
1	1	Robinson, 1983	5	95.00	65.00	64.30	2.83	3.40	5.20	6.60	6.10	3.42	.	1X	530	1
2	1		5	58.00	65.00	64.30	2.83	3.40	5.20	6.60	6.10	4.37	.	8X	530	1
3	1		5	239	65.00	64.30	2.83	3.40	5.20	16.40	15.30	6.36	.	1X	480	1
4	1		5	227	65.00	64.30	2.83	3.40	5.20	16.40	15.30	8.62	.	8X	485	1
5	1		5	188	65.00	64.30	2.83	3.40	4.70	16.10	15.30	8.98	.	2X	485	1
6	1		5	179	65.00	64.30	2.83	3.40	4.80	15.10	14.30	6.83	.	2X	490	1
7	1		5	147	65.00	64.30	2.83	3.40	5.80	15.40	14.60	7.50	.	2X	490	1
8	1		5	135	65.00	64.30	2.83	3.40	5.90	12.70	12.10	6.57	.	2X	494	1
9	1		5	107	65.00	64.30	2.83	3.40	5.10	12.30	11.70	7.59	.	2X	496	1
10	1		5	95.00	65.00	64.30	2.83	3.40	4.60	9.40	8.80	5.59	.	2X	499	1
11	1		5	67.00	65.00	64.30	2.83	3.40	5.10	6.30	5.60	4.11	.	2X	503	1
12	1		5	66.00	65.00	64.30	2.83	3.40	4.10	6.40	5.90	4.16	.	2X	508	1
13	1		5	61.00	65.00	64.30	2.83	3.40	5.40	6.40	6.00	4.39	.	2X	514	1
14	1		5	68.00	68.00	64.30	2.83	3.40	5.10	6.40	6.00	3.98	76.1	2X	490	1
15	1		5	90.00	65.00	64.30	2.83	3.40	5.60	6.40	6.00	4.06	75.3	2X	495	1
16	1		5	85.00	65.00	64.30	2.83	3.40	4.60	6.40	5.90	4.16	79.8	PA	545	1
17	1		5	100	65.00	64.30	2.83	3.40	4.10	6.40	5.90	3.94	81.3	1X	585	1
18	1		5	83.00	65.00	64.30	2.83	3.40	4.10	6.40	5.90	4.21	81.3	4X	555	1
19	1		5	270	65.00	64.30	2.83	3.40	5.00	18.00	16.70	9.18	70.7	4X	501	1
20	1		5	307	65.00	64.30	2.83	3.40	5.00	19.30	18.20	9.32	68.7	PA	517	1
21	1		5	263	65.00	64.30	2.83	3.40	5.00	16.70	15.50	8.98	73.9	1X	508	1
22	1		5	338	65.00	64.30	2.83	3.40	5.10	19.30	18.10	10.69	70.5	1X	490	1
23	1		5	239	65.00	64.30	2.83	3.40	5.10	18.30	17.10	10.57	71.7	4X	502	1
24	1		5	164	65.00	64.30	2.83	3.40	5.00	15.00	14.00	10.50	75.4	PA	510	1
25	1		5	275	65.00	64.30	2.83	3.40	5.00	18.70	17.50	9.25	68.4	PA	513	1
26	1		5	197	65.00	64.30	2.83	3.40	5.60	18.10	16.90	9.22	72.4	1X	522	1
27	1		5	240	65.00	64.30	2.83	3.40	5.60	15.60	14.60	6.66	68.1	4X	531	1
28	2	Tamminga et al., 1979	5	163	40.00	67.30	2.99	4.20	.	8.20	7.70	5.90	80.0	2X	550	1

[a] SPECIES: 1-Sheep, 2-Beef cattle, 5-Dairy cattle
[b] METHODS: 1-DAPA, 2-N15, 3-RNA, 4- Sulfur

OBS	REFNO	SPECIES	BACTYLD	FORAGE	TCN	DE	EE	LIGNIN	DMI	OMI	RDOM	TDOM	FEEDFRQ	BODYWT	METHOD
29	2	5	167	40.00	69.10	3.05	4.20	.	8.20	7.90	6.23	81.0	2X	550	1
30	2	5	154	40.00	69.70	3.07	4.20	.	8.20	7.10	5.76	82.0	2X	550	1
31	2	5	245	31.00	69.80	3.08	4.30	.	12.90	11.60	8.81	79.0	2X	550	1
32	2	5	217	31.00	71.30	3.14	4.30	.	12.90	11.50	9.28	80.0	2X	550	1
33	2	5	244	31.00	72.10	3.18	4.30	.	12.90	12.20	9.89	80.0	2X	550	1
34	3 (Tamminga, 1981b)	5	84.00	40.00	75.00	3.31	4.20	.	7.00	6.50	5.11	.	2X	550	1
35	3	5	187	40.00	75.00	3.31	4.20	.	7.10	6.40	4.92	.	6X	550	1
36	3	5	104	40.00	75.00	3.31	4.20	.	8.40	7.80	6.34	.	2X	550	1
37	3	5	220	40.00	75.00	3.31	4.20	.	8.70	7.80	6.15	.	6X	550	1
38	4 (Tamminga, 1983)	5	176	29.00	79.60	3.51	4.40	.	11.80	10.80	6.35	76.0	2X	550	1
39	4	5	198	40.00	74.00	3.26	4.20	.	12.30	11.20	6.80	77.0	2X	550	1
40	4	5	150	48.00	72.80	3.21	4.00	.	12.10	11.00	6.75	78.0	2X	550	1
41	4	5	139	29.00	74.20	3.27	4.30	.	8.50	7.90	4.58	75.0	2X	550	1
42	4	5	119	56.00	68.80	3.03	3.60	.	8.80	8.20	4.74	76.0	2X	550	1
43	4	5	116	81.00	53.80	2.81	3.00	.	9.20	8.50	4.83	72.0	2X	550	1
44	4	5	261	40.00	72.00	3.17	4.00	.	15.70	14.50	8.49	74.0	2X	550	1
45	4	5	109	31.00	63.80	2.81	3.00	.	7.90	7.10	4.47	78.0	2X	550	1
46	4	5	69.00	42.00	71.60	3.16	4.00	.	5.60	5.20	3.20	80.0	2X	550	1
47	4	5	52.00	83.00	63.40	2.80	3.20	.	3.80	3.50	2.07	75.0	2X	550	1
48	5 (Möller and Hvelplund, 1982)	5	62.00	93.00	49.60	2.19	1.70	10.20	7.00	6.50	4.00	.	2X	775	1
49	5	5	59.00	92.00	48.80	2.15	1.80	10.10	7.10	6.50	4.16	.	2X	775	1
50	5	5	59.00	90.00	48.10	2.12	1.90	9.90	7.10	6.50	3.96	.	2X	775	1
51	5	5	66.00	84.00	53.00	2.34	1.90	9.30	7.10	6.50	4.03	.	2X	775	1
52	5	5	69.00	74.00	56.50	2.49	1.80	8.30	7.00	6.50	4.06	.	2X	775	1
53	6 (Hvelplund and Möller, 1976)	5	68.80	100	56.00	2.47	4.50	3.40	3.50	3.20	2.56	.	2X	410	1
54	6	5	80.50	100	56.00	2.47	4.50	3.40	5.80	5.30	3.64	.	2X	540	1
55	6	5	44.80	100	58.00	2.56	4.30	3.00	3.50	3.20	2.79	.	2X	410	1
56	6	5	72.40	100	58.00	2.56	4.30	3.00	5.80	5.30	3.84	.	2X	540	1

APPENDIX TABLE 3   (*Continued*)

OBS	REFNO	SPECIES	BACTYLD	FORAGE	TDN	DE	EE	LIGNIN	DMI	OMI	RDOM	TDOM	FEEDFRQ	BODYWT	METHOD
57	7 Stern et al., 1983a	5	284	50.00	67.50	2.98	3.30	7.60	17.60	16.70	8.71	61.6	WS	571	1
58	7	5	276	50.00	68.50	3.02	3.20	7.10	17.90	17.00	9.59	66.2	WS	571	1
59	7	5	267	50.00	69.80	3.08	3.10	6.70	17.80	16.90	9.63	69.5	WS	571	1
60	7	5	245	50.00	70.40	3.10	3.00	6.30	17.00	16.20	9.30	71.6	WS	571	1
61	8 Santos et al., 1982	5	328	49.00	65.50	2.89	2.80	8.20	15.30	14.50	7.92	67.1	WS	560	1
62	8	5	333	52.00	64.60	2.85	3.00	8.90	15.10	14.40	7.76	67.6	WS	560	1
63	8	5	302	43.00	68.10	3.00	4.00	9.00	14.00	13.30	6.97	60.7	WS	560	1
64	8	5	308	46.00	67.00	2.95	4.80	9.00	16.00	15.20	7.23	56.7	WS	560	1
65	9 Rode et al., 1983	5	337	25.00	77.50	3.42	2.20	3.90	17.40	16.20	8.32	66.6	4X	582	1
66	9	5	410	75.00	62.50	2.76	2.60	8.00	19.10	17.80	10.85	66.5	4X	582	1
67	9	5	371	25.00	78.30	3.45	3.80	2.30	17.40	16.20	8.63	62.4	4X	582	1
68	9	5	338	75.00	62.80	2.77	3.10	7.80	19.10	17.80	10.00	68.6	4X	582	1
69	10 Rode et al., 1983	5	299	24.00	77.80	3.43	3.60	3.60	16.80	16.00	8.31	68.0	4X	598	1
70	10	5	330	38.00	74.20	3.27	3.40	5.20	17.70	16.70	9.05	67.2	4X	598	1
71	10	5	272	58.00	68.90	3.04	3.30	7.40	16.50	15.40	7.54	59.7	4X	598	1
72	10	5	240	80.00	63.10	2.78	3.10	9.80	14.10	13.10	6.58	56.9	4X	598	1
73	10	5	242	80.00	63.10	2.78	3.10	9.80	14.30	13.20	6.01	53.7	4X	616	1
74	10	5	212	80.00	63.10	2.78	3.10	9.80	13.60	12.60	5.94	54.3	4X	616	1
75	10	5	269	80.00	63.10	2.78	3.10	9.80	14.20	13.10	5.28	54.7	4X	616	1
76	11 Brandt et al., 1981	5	230	64.00	70.20	3.10	3.60	.	13.70	12.80	11.62	73.2	2X	665	2
77	11	5	227	64.00	68.40	3.02	3.50	.	14.20	13.30	11.54	70.1	2X	665	2
78	11	5	227	64.00	68.40	3.02	3.50	.	14.10	13.20	11.50	71.0	2X	665	2
79	11	5	203	64.00	69.90	3.08	3.60	.	13.60	12.80	10.28	66.0	6X	665	2
80	12 Hagemeister and Pfeffer, 1973	5	171	34.60	72.10	3.18	.	.	9.50	8.60	6.23	79.2		.	1
81	12	5	107	34.60	72.10	3.18	.	.	7.50	6.80	4.66	77.1		.	1
82	12	5	106	34.60	72.10	3.18	.	.	9.30	8.40	4.72	67.8		.	1
83	12	5	102	34.60	72.10	3.13	.	.	7.80	7.00	3.96	74.1		.	1
84	12	5	184	34.60	72.10	3.13	.	.	9.60	8.60	6.09	69.2		.	1

OBS	REFNO	SPECIES	BACTYLD	FORAGE	TDN	DE	EE	LIGNIN	DMI	OMI	RDOM	TDOM	FEEDFRQ	BODYWT	METHOD
85	12	5	162	34.60	72.10	3.18	.	.	7.80	7.00	4.62	69.2		.	1
86	12	5	181	35.80	69.30	3.0o	.	.	9.90	8.90	6.01	76.6		.	1
87	12	5	127	35.30	69.30	3.06	.	.	7.90	7.10	4.50	75.3		.	1
88	12	5	150	33.00	72.10	3.18	.	.	9.50	8.60	6.15	72.7		.	1
89	12	5	107	33.00	72.10	3.18	.	.	7.60	6.80	4.33	71.5		.	1
90	12	5	149	34.90	69.50	3.06	.	.	9.60	8.60	6.12	74.3		.	1
91	12	5	136	34.90	69.50	3.06	.	.	8.30	7.50	5.21	76.8		.	1
92	13 Hagemeister and Kaufmann, 1974	5	79.30	100	55.00	2.42	.	.	7.30	6.60	.	61.5		.	1
93	13	5	69.00	100	55.00	2.42	.	.	7.40	6.70	.	62.5		.	1
94	13	5	76.00	100	60.00	2.65	.	.	8.20	7.40	.	68.8		.	1
95	13	5	105	34.40	72.20	3.18	.	.	8.60	7.70	.	70.9		.	1
96	13	5	85.00	41.40	68.00	3.00	.	.	9.10	8.20	.	74.1		.	1
97	13	5	134	35.80	72.00	3.17	.	.	8.90	8.00	.	75.2		.	1
98	13	5	153	41.60	63.80	3.03	.	.	9.20	8.30	.	72.9		.	1
99	13	5	149	32.90	71.20	3.14	.	.	9.90	8.90	.	70.1		.	1
100	13	5	139	42.40	67.30	2.97	.	.	9.20	8.30	.	75.7		.	1
101	13	5	133	42.40	68.70	3.03	.	.	9.20	8.30	.	74.9		.	1
102	13	5	129	47.40	67.30	2.97	.	.	9.50	8.60	.	76.1		.	1
103	13	5	137	34.70	72.20	3.18	.	.	8.50	7.70	.	77.9		.	1
104	13	5	144	35.30	69.70	3.07	.	.	8.80	7.90	.	75.2		.	1
105	14 Oldham et al., 1979	5	216	40.00	76.00	3.35	.	.	13.50	12.20	8.43	.		590	1
106	14	5	136	10.00	76.00	3.35	.	.	12.60	11.30	7.49	.		590	1
107	14	5	194	40.00	71.50	3.15	.	.	13.50	12.20	8.15	.		590	1
108	14	5	70.00	10.00	71.50	3.15	.	.	12.60	11.30	5.96	.		590	1
109	15 Arambel and Coon, 1981	5	8.80	100	40.60	1.79	5.20	7.50	4.80	4.30	.	.	2X	614	3
110	15	5	57.10	43.00	63.30	2.79	5.50	3.80	11.30	10.20	.	.	2X	614	3
111	15	5	104.2	43.00	61.90	2.73	5.50	3.80	11.30	10.20	.	.	2X	614	3
112	15	5	73.40	43.00	63.30	2.79	5.30	4.00	11.30	10.20	.	.	2X	614	3

APPENDIX TABLE 3   (*Continued*)

OBS	REFNO	SPECIES	BACTYLD	FORAGE	TDN	DE	EE	LIGNIN	DMI	OMI	RDOM	TDOM	FEEDFRQ	BODYWT	METHOD
113	15	5	76.40	43.00	63.60	2.80	4.70	3.80	11.30	10.20	.	.	2X	614	3
114	15	5	31.30	43.00	58.70	2.59	6.10	4.60	11.30	10.20	.	.	2X	614	3
115	15	5	60.00	43.00	62.50	2.76	4.90	3.90	11.30	10.20	.	.	2X	614	3
116	49 Merchen and Satter, 1983	5	257	65.00	61.20	2.70	2.30	9.40	17.60	16.30	10.51	69.8	4X	595	1
117	49	5	314	65.00	62.50	2.76	2.60	8.10	20.00	18.50	12.47	73.6	4X	595	1
118	49	5	287	65.00	62.50	2.76	2.60	8.10	19.00	17.60	10.02	71.5	4X	595	1
119	49	5	229	65.00	58.60	2.58	2.30	10.40	17.90	16.50	9.57	68.0	4X	595	1
120	16 Beever et al., 1981	1	14.00	100	70.00	3.09	.	2.80	1.04	0.94	0.64	79.4	2X	50	4
121	16	1	12.90	100	70.00	3.09	.	3.20	1.08	0.98	0.53	74.4	2X	50	4
122	16	1	13.90	100	60.00	2.65	.	3.10	1.03	0.93	0.60	74.3	2X	50	4
123	16	1	12.00	100	60.00	2.65	.	4.20	1.07	0.97	0.53	69.0	2X	50	4
124	17 Chamberlain and Thomas, 1979	1	8.60	100	60.00	2.65	.	.	0.58	0.53	0.38	70.0	2X	45	1
125	17	1	9.80	71.00	67.50	2.98	.	.	0.59	0.56	0.38	77.0	2X	45	1
126	17	1	10.10	43.00	74.20	3.27	.	.	0.60	0.57	0.43	82.0	2X	45	1
127	17	1	7.70	14.00	81.70	3.60	.	.	0.60	0.59	0.47	90.0	2X	45	1
128	17	1	6.60	0.00	85.00	3.75	.	.	0.60	0.59	0.46	91.0	2X	45	1
129	18 Chamberlain and Thomas, 1980b	1	10.50	18.00	81.00	3.57	.	.	0.73	0.69	0.51	.	2X	52	1
130	18	1	12.10	18.00	81.00	3.57	.	.	0.73	0.70	0.52	.	2X	52	1
131	18	1	10.70	18.00	81.00	3.57	.	.	0.73	0.69	0.52	.	2X	52	1
132	18	1	13.10	18.00	81.00	3.57	.	.	0.73	0.70	0.50	.	2X	52	1
133	19 Chamberlain and Thomas, 1976	1	5.30	0.00	75.60	3.33	.	.	0.35	0.33	0.27	88.9	4X	26	1
134	19	1	3.90	0.00	75.60	3.33	.	.	0.34	0.32	0.28	81.0	4X	26	1
135	19	1	6.80	0.00	75.60	3.33	.	.	0.44	0.42	0.35	90.4	4X	26	1
136	19	1	9.20	0.00	75.60	3.33	.	.	0.59	0.55	0.46	89.9	4X	26	1
137	19	1	10.00	0.00	75.60	3.33	.	.	0.67	0.63	0.53	88.2	4X	45	1
138	19	1	9.90	0.00	75.60	3.33	.	.	0.67	0.63	0.52	91.0	4X	45	1
139	19	1	11.50	0.00	75.60	3.33	.	.	0.70	0.65	0.55	89.6	4X	45	1
140	20	1	14.30	49.60	66.50	2.93	.	4.80	0.81	0.76	0.52	80.9	12X	.	4

OBS	REFNO	SPECIES	BACTYLD	FORAGE	TDN	DE	EE	LIGNIN	DMI	OMI	RDOM	TDOM	FEEDFRQ	BODYWT	METHOD
141	20 Hume, 1970	1	16.20	49.90	66.50	2.93	.	6.40	0.81	0.76	0.43	80.2	12X	.	4
142	20	1	14.50	50.00	66.50	2.93	.	6.80	0.80	0.75	0.46	79.7	12X	.	4
143	20	1	16.70	49.90	66.50	2.93	.	6.30	0.81	0.77	0.46	80.7	12X	.	4
144	21 Hume and Purser, 1974	1	16.00	100	75.00	3.31	.	4.80	0.90	0.80	0.69	80.6	12X	55	4
145	21	1	8.40	100	55.00	2.42	.	6.40	0.74	0.66	0.25	63.6	12X	55	4
146	21	1	7.30	100	55.00	2.42	.	6.80	0.70	0.62	0.29	59.0	12X	55	4
147	21	1	8.40	100	58.00	2.56	.	6.30	0.70	0.66	0.39	64.9	12X	55	4
148	22 Kennedy et al., 1976	1	23.60	100	56.00	2.47	.	.	1.84	1.64	0.72	51.0	24X	54	4
149	22	1	21.40	100	56.00	2.47	.	.	1.84	1.64	0.62	48.0	24X	52	4
150	23 Kennedy et al., 1981	1	21.90	100	56.00	2.47	.	.	1.41	1.31	0.76	58.0	24X	59	4
151	23	1	20.80	100	56.00	2.47	.	.	1.41	1.31	0.59	53.0	24X	51	4
152	23	1	29.80	100	56.00	2.47	.	.	2.35	2.18	0.83	45.0	24X	57	4
153	24 Mathers and Miller, 1981	1	10.30	100	63.40	2.80	2.80	8.50	0.69	0.65	0.35	.	12X	30	4
154	24	1	12.10	66.70	68.80	3.03	2.50	6.30	0.69	0.66	0.40	.	12X	30	4
155	24	1	10.70	33.30	74.10	3.27	2.20	4.10	0.70	0.66	0.47	.	12X	30	4
156	24	1	13.10	0.00	79.40	3.50	1.90	1.90	0.70	0.67	0.52	.	12X	30	4
157	25 Leibolz and Hartmann, 1972	1	4.50	98.00	62.50	2.76	3.30	6.00	0.74	0.64	0.43	75.3	24X	50	4
158	25	1	4.80	50.00	73.90	3.26	1.40	1.00	0.48	0.45	0.34	59.0	24X	50	4
159	25	1	5.30	36.00	65.30	2.88	2.90	5.50	0.79	0.72	0.54	73.9	24X	50	4
160	25	1	8.00	50.00	73.10	3.22	1.50	1.00	0.78	0.73	0.50	71.9	24X	50	4
161	25	1	9.00	50.00	73.30	3.23	1.50	1.00	0.80	0.72	0.51	75.1	24X	50	4
162	25	1	4.70	83.00	54.60	2.41	2.10	1.70	0.71	0.64	0.43	65.7	24X	50	4
163	26 Mercer et al., 1980	1	11.70	10.00	76.00	3.35	1.70	2.90	0.64	0.60	0.47	76.1	24X	40	1
164	26	1	12.10	10.00	76.70	3.38	1.80	2.90	0.63	0.59	0.49	77.3	24X	40	1
165	26	1	9.90	10.00	76.90	3.39	2.50	2.80	0.63	0.60	0.40	76.3	24X	40	1
166	27 Sutton et al., 1975	1	10.00	16.00	77.30	3.41	1.80	1.90	0.55	0.53	0.41	81.3	2X	40	3
167	27	1	9.00	16.00	76.40	3.37	1.70	1.70	0.55	0.53	0.42	82.9	2X	40	3
168	27	1	10.00	16.00	75.30	3.32	1.80	1.70	0.55	0.53	0.41	81.5	2X	40	3

OBS	REFNO	SPECIES	BACTYLD	FORAGE	TDN	DE	EE	LIGNIN	DMI	OMI	RDOM	TDOM	FEEDFRQ	BODYWT	METHOD
169	27	1	8.30	50.00	67.50	2.98	1.80	1.90	0.50	0.47	0.30	68.0	2X	40	3
170	27	1	8.70	50.00	71.10	3.13	4.80	1.80	0.55	0.52	0.32	68.5	2X	40	3
171	28 (Ulyatt et al., 1975b)	1	6.90	100	68.00	3.00	3.00	2.00	0.54	0.50	0.31	.	2X	42	1
172	28	1	10.40	100	68.00	3.00	3.00	2.00	0.86	0.80	0.54	.	2X	42	1
173	28	1	11.70	100	72.00	3.17	3.00	2.00	0.54	0.50	0.33	.	2X	42	1
174	28	1	16.20	100	72.00	3.17	3.00	2.00	0.86	0.80	0.52	.	2X	42	1
175	28	1	3.30	100	75.00	3.31	2.70	6.00	0.54	0.50	0.35	.	2X	42	1
176	28	1	14.10	100	75.00	3.31	2.70	6.00	.	0.80	0.56	.	2X	42	1
177	29 (Walker et al., 1975)	1	13.90	100	70.00	3.09	4.00	5.00	0.91	0.86	0.64	75.5	24X	40	4
178	29	1	14.60	100	70.00	3.09	4.00	5.00	0.91	0.86	0.54	75.5	24X	40	4
179	29	1	10.00	100	70.00	3.09	4.00	5.00	0.91	0.36	0.44	75.5	24X	40	4
180	29	1	8.70	100	70.00	3.09	3.00	5.00	0.89	0.85	0.65	83.5	24X	40	4
181	29	1	10.50	100	70.00	3.09	3.00	7.00	0.89	0.85	0.66	83.5	24X	40	4
182	29	1	13.60	100	70.00	3.09	3.00	7.00	0.89	0.85	0.66	83.5	24X	40	4
183	29	1	9.20	100	58.00	2.56	2.20	7.00	0.90	0.85	0.55	66.0	24X	40	4
184	29	1	7.00	100	58.00	2.56	2.20	7.00	0.90	0.85	0.43	66.0	24X	40	4
185	29	1	7.70	100	58.00	2.56	2.20	7.00	0.90	0.85	0.42	66.0	24X	40	4
186	29	1	5.30	100	58.00	2.56	2.00	10.50	0.64	0.60	0.30	63.5	24X	40	4
187	29	1	6.00	100	58.00	2.56	2.00	10.50	0.82	0.78	0.32	63.5	24X	40	4
188	48 (Merchen and Satter, 1983)	1	8.80	100	63.00	2.73	3.80	7.90	0.59	0.54	0.32	.	12X	25	1
189	48	1	8.40	100	63.00	2.78	3.80	7.00	0.60	0.53	0.35	.	12X	25	1
190	30 (Cole et al., 1976b)	2	34.00	0.00	85.90	3.79	4.00	1.90	4.46	4.29	3.12	55.0	24X	465	3
191	30	2	40.30	7.00	75.30	3.32	3.80	3.60	5.23	4.93	3.06	52.0	24X	465	3
192	30	2	47.10	14.00	72.20	3.18	3.60	5.40	5.69	5.43	3.30	46.0	24X	465	3
193	30	2	57.40	21.00	69.10	3.05	3.40	7.10	5.93	5.63	3.46	49.0	24X	465	3
194	30	2	32.50	10.50	77.50	3.42	3.70	4.50	4.84	4.66	3.57	.	24X	465	3
195	30	2	37.00	10.50	77.50	3.42	3.70	4.50	4.85	4.66	2.87	.	24X	465	3
196	30	2	24.00	0.00	85.90	3.79	4.00	1.90	4.36	4.19	3.08	.	24X	465	3

OBS	REFNO	SPECIES	BACTYLD	FORAGE	TDN	DE	EE	LIGNIN	DMI	OMI	RDOM	TDOM	FEEDFRQ	BODYWT	METHOD
197	30	2	45.40	21.00	69.10	3.05	3.40	7.10	5.43	5.12	3.32	.	24X	465	3
198	31 Kropp et al., 1977b	2	57.00	77.30	53.00	2.34	1.70	17.30	4.52	4.31	2.13	60.5	24X	350	3
199	31	2	53.60	75.80	53.20	2.35	1.90	17.20	4.51	4.33	2.05	51.6	24X	350	3
200	31	2	54.50	74.90	52.90	2.33	1.90	17.10	4.50	4.31	1.93	56.0	24X	350	3
201	31	2	55.70	74.60	52.40	2.31	1.80	17.00	4.50	4.29	2.02	58.7	24X	350	3
202	32 Merchen et al., 1979	2	64.10	60.00	61.10	2.69	0.43	4.20	8.32	7.50	5.41	.	24X	262	1
203	32	2	66.10	62.50	60.90	2.69	0.56	4.50	8.41	7.50	5.19	.	24X	262	1
204	32	2	54.70	53.10	63.30	2.79	1.60	4.70	8.31	7.52	5.08	.	24X	262	1
205	33 Prigge et al., 1978	2	29.20	15.00	77.80	3.43	3.40	3.90	4.40	3.88	2.72	80.0	2X	429	3
206	33	2	35.40	15.00	77.80	3.43	3.40	4.60	4.49	4.37	2.71	82.0	2X	429	3
207	33	2	35.20	15.00	77.80	3.43	3.40	4.40	4.44	4.30	2.67	81.0	2X	429	3
208	33	2	36.80	15.00	77.80	3.43	3.40	4.40	4.51	4.37	1.92	82.0	2X	429	3
209	34 Zinn and Owens, 1983	2	35.00	20.00	77.20	3.40	3.70	1.30	3.81	3.61	1.84	79.0	2X	203	3
210	34	2	55.00	20.00	76.90	3.39	3.20	2.50	4.79	4.52	2.12	75.0	2X	203	3
211	34	2	47.00	20.00	75.20	3.32	3.80	2.50	4.77	4.47	2.47	77.0	2X	203	3
212	34	2	51.00	20.00	79.20	3.49	3.10	1.30	4.75	4.49	2.32	80.0	2X	203	3
213	34	2	44.00	20.00	75.20	3.32	3.70	1.30	4.11	3.90	2.10	76.0	2X	227	3
214	34	2	55.00	20.00	76.20	3.36	3.80	2.50	5.05	4.73	2.29	68.0	2X	227	3
215	34	2	58.00	20.00	78.40	3.46	3.40	1.20	4.57	4.25	1.93	76.0	2X	227	3
216	34	2	51.00	20.00	76.00	3.35	4.90	1.10	5.00	4.70	2.12	75.0	2X	227	3
217	34	2	44.00	60.00	61.70	2.72	2.20	3.70	4.67	4.39	2.10	67.0	2X	254	3
218	34	2	51.00	60.00	63.10	2.78	2.10	4.00	4.70	4.42	1.59	62.0	2X	254	3
219	34 McAllan and Smith, 1983	2	44.00	60.00	64.20	2.83	2.10	3.40	4.66	4.38	2.09	68.0	2X	254	3
220	35	2	31.60	44.90	68.40	3.02	1.90	5.50	3.34	3.30	2.28	.	2X	193	1
221	35	2	49.20	43.90	69.00	3.04	1.90	5.40	3.42	3.14	2.01	.	2X	193	1
222	35	2	57.60	43.90	69.00	3.04	1.90	5.40	3.42	3.21	2.34	.	2X	193	1
223	35	2	59.30	45.00	68.40	3.02	1.90	5.50	4.89	4.52	2.89	.	2X	193	1
224	35	2	59.60	43.90	69.00	3.04	1.90	5.40	5.01	4.55	2.82	.	2X	193	1

APPENDIX TABLE 3 (*Continued*)

OBS	REFNO	SPECIES	BACTYLD	FORAGE	TDN	DE	EE	LIGNIN	DMI	OMI	RDOM	TDOM	FEEDFRQ	BODYWT	METHOD
225	35	2	63.10	43.90	69.00	3.04	1.90	5.40	5.01	4.66	2.75	.	2X	193	1
226	36 Redman et al., 1980	2	27.00	19.70	72.80	3.21	1.90	2.30	3.80	3.14	1.75	92.0	8X	150	1
227	36	2	31.80	17.50	65.20	2.87	1.60	1.90	3.78	3.16	2.13	59.0	8X	150	1
228	36	2	29.10	19.70	73.50	3.24	1.60	2.40	3.81	3.14	2.05	58.0	8X	150	1
229	36	2	24.40	18.30	74.80	3.30	1.70	2.20	3.83	3.16	1.86	63.0	8X	150	1
230	36	2	24.00	18.00	75.90	3.35	1.60	2.20	3.77	3.19	1.98	62.0	8X	150	1
231	37 Thomson et al., 1981	3	24.10	100	68.00	3.00	2.50	2.00	2.51	2.30	1.51	76.4	2X	141	4
232	37	3	27.90	100	68.00	3.00	2.50	2.00	2.65	2.43	1.49	73.2	2X	141	4
233	37	3	23.30	100	68.00	3.00	2.50	2.00	2.34	2.15	1.35	74.5	2X	141	4
234	38 Veira et al., 1981	3	33.30	7.50	78.50	3.46	3.90	1.20	3.44	3.30	2.03	.	2X	146	1
235	38	3	29.20	7.50	76.10	3.36	3.90	1.30	3.52	3.36	1.86	.	2X	146	1
236	38	3	34.60	7.50	76.40	3.37	3.70	1.30	3.50	3.34	1.76	.	2X	146	1
237	38	3	29.00	7.50	76.80	3.39	3.50	1.40	3.50	3.33	1.95	.	2X	146	1
238	39 Zinn and Owens, 1980	2	27.80	20.00	64.20	2.83	3.50	1.40	3.01	2.85	1.61	81.0	2X	232	3
239	39	2	41.30	20.00	64.20	2.83	3.50	1.40	4.62	4.37	2.09	78.0	2X	232	3
240	39	2	43.50	40.00	70.80	3.12	3.00	2.60	4.11	3.85	2.06	71.0	2X	232	3
241	40 Kropp et al., 1977a	2	30.60	75.30	59.80	2.64	2.20	7.70	4.61	4.24	1.79	55.0	24X	275	3
242	40	2	32.50	75.40	58.20	2.57	2.30	7.70	4.52	4.25	1.84	52.0	24X	275	3
243	40	2	31.00	75.40	62.20	2.74	2.40	7.70	4.57	4.19	1.70	46.0	24X	275	3
244	40	2	33.00	75.40	58.00	2.56	2.40	7.50	4.59	4.23	1.75	49.0	24X	275	3
245	41	2	46.40	10.50	78.30	3.45	3.70	4.40	4.84	4.70	3.71	84.0	24X	460	.
246	41	2	44.60	10.50	69.20	3.05	3.70	4.40	4.85	4.70	3.01	78.0	24X	460	.
247	41	2	42.60	0.00	78.30	3.45	4.00	1.90	4.36	4.20	3.28	85.0	24X	460	.
248	41	2	48.50	21.60	69.20	3.05	3.40	7.10	5.43	5.10	3.31	76.0	24X	460	.
249	42 Theuer, 1969	2	74.70	.	.	.	.	.	.	6.30	4.72	78.0		410	.
250	42	2	31.30	.	.	.	.	.	.	6.30	5.53	78.0		410	.
251	43 Isichei, 1980	2	82.90	.	.	.	.	.	.	6.40	2.82	75.0		288	.
252	43	2	74.90	.	.	.	.	.	.	6.20	2.98	82.0		288	.

OBS	REFNO	SPECIES	BACTYLD	FORAGE	TDN	DE	EE	LIGNIN	DMI	OMI	RCOM	TDOM	FEEDFRQ	BODYWT	METHOD
253	43	2	95.50	.	.	.	.	.	.	6.80	2.99	70.0		288	.
254	43	2	76.20	.	.	.	.	.	.	6.50	3.38	69.0		288	.
255	44 Poos et al., 1979	2	69.40	0.00	65.10	2.87	2.40	4.20	.	7.00	4.69	65.0	24X	262	1
256	44	2	47.40	0.00	64.10	2.83	2.40	4.20	.	7.00	4.13	60.0	24X	262	1
257	44	2	84.20	0.00	65.10	2.87	1.30	3.70	.	7.00	4.97	67.0	24X	262	1
258	44	2	56.20	0.00	64.10	2.83	1.30	3.70	.	7.00	4.97	64.0	24X	262	1
259	46 Zinn and Owens, 1980	2	44.00	.	.	.	.	.	4.50	4.30	2.33	87.0	16X	288	.
260	46	2	49.00	.	.	.	.	.	4.70	4.40	2.08	74.1	15X	327	.
261	46	?	42.00	.	.	.	.	.	4.70	4.30	1.84	61.8	9X	251	.
262	46	2	38.00	.	.	.	.	.	4.80	4.40	1.92	65.6	9X	254	.
263	46	2	47.00	.	.	.	.	.	6.00	5.60	2.42	65.7	9X	375	.
264	46	2	49.00	.	.	.	.	.	5.10	4.80	2.36	87.3	9X	282	.
265	46	2	42.00	.	.	.	.	.	4.90	4.60	2.26	76.9	8X	248	.
266	46	2	41.00	.	.	.	.	.	4.50	4.10	2.44	78.2	11X	217	.
267	46	2	39.00	.	.	.	.	.	3.70	3.50	1.85	76.7	14X	259	.
268	46	2	48.00	.	.	.	.	.	3.70	3.50	1.52	58.8	6X	348	.
269	46	2	37.00	.	.	.	.	.	4.40	4.10	1.78	53.4	3X	348	.
270	46	2	68.00	.	.	.	.	.	6.80	6.50	3.68	76.0	16X	341	.
271	47 Weakley et al., 1983	2	.	.	.	.	.	.	.	3.40	2.39	79.3		254	.
272	47	2	.	.	.	.	.	.	.	3.40	2.51	84.9		254	.
273	47	2	.	.	.	.	.	.	.	3.40	1.78	68.6		254	.
274	47	2	.	.	.	.	.	.	.	3.40	2.22	73.4		254	.
275	47	2	.	.	.	.	.	.	.	5.00	3.19	77.1		490	.
276	47	2	.	.	.	.	.	.	.	5.40	2.96	78.2		490	.
277	47	2	.	.	.	.	.	.	.	5.30	3.16	79.9		490	.
278	47	2	.	.	.	.	.	.	.	5.20	3.19	73.7		490	.
279	47	2	.	.	.	.	.	.	.	5.60	3.86	83.4		490	.
280	47	2	.	.	.	.	.	.	.	5.50	3.89	84.8		490	.

Description of Appendix Tables 4 to 6

Appendix Tables 4 through 6 present dietary protein levels (IPDM) above which added nonprotein nitrogen (NPN) sources such as urea should not be useful. Estimates are provided for various concentrations of dietary energy (TDN) and of feed intake (TDNI or percentage of body weight) and various degrees of ruminal proteolysis of dietary protein (DIPIP). Values for protein use in the rumen for each species were calculated from regression equations that relate synthesis of microbial protein (BCP) to energy intake. Ruminal ammonia supply was calculated from extent of proteolysis (DIPIP) and recycling (RP). A maximum efficiency of 0.90 was used for ammonia capture by ruminal microbes.

Utilization of ruminally degraded protein (DIP) was assumed to equal the synthesis of bacterial crude protein (BCP) calculated from regression equations relating bacterial protein yield to TDN intake for dairy cattle and for sheep as discussed in the chapter concerning microbial activities in the rumen. Values for BCP for beef cattle additionally considered the changes in efficiency of microbial efficiency with level of roughage (assumed 55 percent TDN) and concentrate (assumed 90 percent TDN) in the diet and intake expressed as a percentage of body weight. Efficiency of conversion of DIP to BCP was assumed to be 0.90 though greater efficiency might be expected with lower ruminal ammonia concentrations.

Recycling of nitrogen to the rumen (RP) was considered to equal 0.15 IP for all species in these tables. For animals fed very low protein diets, RP might be more accurately predicted from dietary protein percentage rather than by assuming that only 15 percent of IP can be recycled. This is described in the text chapter. Further information concerning RP with various dietary concentrate levels and IPDM levels is needed to more precisely predict dietary conditions to which nonprotein nitrogen additions will be useful.

Protein values in Appendix Tables 4, 5, and 6 indicate that urea can be added successfully to many diets for dairy cattle, beef cattle, and sheep. The level of IP at which urea addition is useful increases as extent of IP-resisting degradation in the rumen is increased (UIPIP) for all species. An increase in UIPIP decreases the amount of ammonia available for BCP synthesis so NPN can be included as a substitute source of ammonia.

The level of IP at which urea addition is useful will increase as dietary TDN concentration increases for both cattle and sheep. This is a result of the increase in BCP synthesis that is driven by TDN concentration of the diet and feed intake. A similar relationship was not apparent in the values calculated for beef cattle. This was because efficiency of conversion of TDN to BCP dropped as concentrate level increased so that urea usefulness declined at the higher energy levels. Lower efficiency of BCP synthesis at higher concentrate levels is apparent from data from other species as well, though few experiments with high concentrate levels for other species are available.

Finally, TDN intake influenced urea usefulness for dairy cattle and sheep. Values changed because the regression equation for calculating BCP synthesis has an intercept value that is not zero. For beef cattle, urea usefulness also changed with feed intake level. This is a result of the influence of feed intake level on efficiency of BCP synthesis. Higher efficiencies of BCP synthesis have been reported at higher feed intake levels from a number of trials.

As these estimates for urea utilization have been cal-

APPENDIX TABLE 4   Urea Usefulness with Various Feed Intakes, TDN Levels, and Ruminal Digestions of Dietary Protein Based on Equations from Dairy Cattle

	Percent Dietary Protein Escaping Digestion in the Rumen											
	20	20	20	30	30	30	40	40	40	50	50	50
	Daily TDN Intake, kg.											
	5	10	15	5	10	15	5	10	15	5	10	15
TDN%	Percent Dietary Protein above which Urea is useless[a]											
55	7.94	9.22	9.65	8.87	10.31	10.78	10.06	11.68	12.22	11.60	13.48	14.10
60	8.66	10.06	10.52	9.68	11.24	11.76	10.97	12.74	13.33	12.66	14.70	15.38
65	9.38	10.90	11.40	10.49	12.18	12.74	11.89	13.80	14.44	13.71	15.93	16.66
70	10.10	11.74	12.28	11.29	13.12	13.72	12.80	14.86	15.55	14.77	17.15	17.95
75	10.83	12.57	13.16	12.10	14.05	14.70	13.71	15.93	16.66	15.82	18.38	19.23
80	11.55	13.41	14.03	12.91	14.99	15.68	14.63	16.99	17.77	16.88	19.60	20.51

[a]Calculated as: [(26.12*TDNI)-31.86)*6.25]/[.9*(115-%UIP)*10*(TDNI/TDN%)], where TDNI is TDN intake, UIP is percent escape protein and TDN% is TDN percentage.

APPENDIX TABLE 5   Urea Usefulness with Various Feed Intakes, TDN Levels, and Ruminal Digestions of Dietary Protein Based on Equations from Beef Cattle

	Percent Dietary Protein Escaping Digestion in the Rumen											
	20	20	20	30	30	30	40	40	40	50	50	50
	Daily Intake, % of body weight											
	1.75	2.00	2.25	1.75	2.00	2.25	1.75	2.00	2.25	1.75	2.00	2.25
TDN%	Percent Dietary Protein above which Urea is useless[a]											
75	9.46	9.88	10.23	10.58	11.04	11.43	11.99	12.51	12.96	13.83	14.44	14.95
80	8.99	9.43	9.84	10.05	10.54	11.00	11.39	11.95	12.47	13.14	13.79	14.39
85	7.98	8.33	8.66	8.92	9.31	9.68	10.11	10.55	10.97	11.67	12.17	12.66
90	6.36	6.46	6.56	7.11	7.22	7.33	8.06	8.18	8.31	9.30	9.44	9.59

[a]Calculated as: [DMI*TDN%*6.25*{(8.63+14.60*DMI*6.25*((90-TDN)/35)-5.18*DMI*DMI*((90-TDN)^2/(35*35))+.5953*DMI*((TDN-55)/35)}]/[.9*(115-%UIP)*10*DMI], where DMI is DMI intake as % of body weight, UIP is percent escape protein and TDN% is TDN percentage.

APPENDIX TABLE 6   Urea Usefulness with Various Feed Intakes, TDN Levels, and Ruminal Digestions of Dietary Protein Based on Equations from Sheep

	Percent Dietary Protein Escaping Digestion in the Rumen											
	20	20	20	30	30	30	40	40	40	50	50	50
	Daily TDN Intake, kg.											
	.5	1	1.5	.5	1	1.5	.5	1	1.5	.5	1	1.5
TDN%	Percent Dietary Protein above which Urea is useless[a]											
55	8.23	8.74	8.92	9.19	9.77	9.97	10.42	11.08	11.30	12.02	12.78	13.03
60	8.97	9.54	9.73	10.03	10.66	10.87	11.37	12.08	12.32	13.12	13.94	14.22
65	9.72	10.33	10.54	10.87	11.55	11.78	12.31	13.09	13.35	14.21	15.10	15.40
70	10.47	11.13	11.35	11.70	12.44	12.68	13.26	14.10	14.38	15.30	16.27	16.59
75	11.22	11.92	12.16	12.54	13.33	13.59	14.21	15.10	15.40	16.39	17.43	17.77
80	11.96	12.72	12.97	13.37	14.22	14.50	15.16	16.11	16.43	17.49	18.59	18.96
85	12.71	13.51	13.78	14.21	15.10	15.40	16.10	17.12	17.46	18.58	19.75	20.14

[a]Calculated as: [((23.04*TDNI)-1.29)*6.25]/[.9*(115-%UIP)*10*(TDNI/TDN%)], where TDNI is TDN intake, UIP is percent escape protein and TDN% is TDN percentage.

culated based on a number of equations, all of which have not been verified, application of the values cannot be recommended under all feeding conditions. However, these values should serve as guides to indicate the general conditions under which urea can be included in a diet as a substitute for other protein sources. If these values indicate that urea can be used, and if urea substitution for other protein sources will reduce feed cost, it should be included in the diet.

Description of Appendix Tables 7 and 8

Protein requirements for various types of growing-finishing beef cattle are presented in Appendix Tables 7 and 8. These were calculated from equations cited in various chapters and based on feed intakes and performance enumerated by NRC (1984). A sample calculation is provided below.

A 300-kg (shrunk weight) medium-frame steer gaining at the rate of 1.2 kg (shrunk weight) daily requires 4.9 Mcal of NE$_g$. To consume enough feeds to achieve this intake of NE$_g$, the diet must contain a minimum of 2.74 Mcal metabolizable energy per kg of dry feed. This value was obtained by iterating feed intake and gain equations (NRC, 1984) at various ME levels. These iterated values for various rates of gain are provided in Appendix Table 7. At the determined energy intake (7.2 kg of feed at 75.9 percent TDN), a typical feed would contain 59.7 percent concentrate and 40.3 percent forage if concentrate and forage were assumed to provide 55 and 90 percent TDN, respectively. (Actual values for roughage and concentrate could be used in the field.) With this feed intake level (7.2 kg per 300 kg body weight), this steer is consuming a total of 2.06 percent

APPENDIX TABLE 7   Production of Microbial Protein and Escape Protein Requirements for Beef Cattle of Various Weights and Types

Microbial Protein Production, g/day

Gain kg/day	ME mcal/kg	150	200	250	300	350	400	450	500	550	600
Medium frame steer											
0.20	1.94	143	212	275	335	392	446	499	549	598	
0.40	2.08	189	267	339	407	471	532	591	648	703	
0.60	2.22	247	333	412	486	556	623	687	749	809	
0.80	2.36	309	399	483	561	635	705	773	838	901	
1.00	2.53	370	462	548	627	702	773	842	908	971	
1.20	2.74	398	485	566	641	711	778	842	904	964	
Large frame steer											
0.20	1.88	103	172	237	298	356	411	465	517	567	616
0.40	2.00	137	215	288	357	423	486	546	605	661	716
0.60	2.11	182	268	348	424	496	564	630	694	756	816
0.80	2.23	242	336	423	504	581	655	726	794	860	925
1.00	2.35	307	406	497	583	664	741	815	887	956	1023
1.20	2.49	374	477	570	658	741	820	895	968	1038	1106
1.40	2.65	424	524	616	702	783	860	934	1005	1073	1139
1.60	2.87	419	507	589	664	736	804	869	931	992	1051
Medium frame bull											
0.20	1.92	138	205	267	326	382	435	486	536	583	630
0.40	2.05	178	254	324	390	453	513	571	626	680	733
0.60	2.17	226	309	385	457	526	591	653	714	772	829
0.80	2.30	283	372	453	530	603	672	738	802	864	924
1.00	2.44	340	433	518	597	672	743	812	877	941	1002
1.20	2.60	386	478	562	641	715	786	854	919	981	1042
1.40	2.82	391	474	550	622	689	753	814	873	930	985
Large frame bull											
0.20	1.89	120	187	250	309	365	419	471	521	570	617
0.40	2.00	150	226	296	362	425	485	543	599	653	706
0.60	2.10	188	270	347	419	487	552	615	676	734	791
0.80	2.21	238	327	409	486	560	629	696	761	824	884
1.00	2.32	292	386	472	553	630	703	773	841	906	969
1.20	2.45	351	448	537	621	699	775	846	916	983	1047
1.40	2.58	395	492	581	664	743	817	889	957	1024	1088
1.60	2.76	415	506	590	668	741	811	878	942	1004	1064
1.80	3.07	321	388	450	507	562	614	665	713	760	806
Medium frame heifer											
0.20	2.01	179	246	307	365	420	473	523	572	619	
0.40	2.20	242	319	389	456	518	578	635	690	744	
0.60	2.39	308	390	466	537	604	667	728	787	844	
0.80	2.62	358	440	516	586	652	715	775	833	889	
1.00	3.15	237	288	335	380	422	463	502	540	576	
Large frame heifer											
0.20	1.94	143	212	275	335	392	446	499	549	598	645
0.40	2.09	193	271	344	412	477	539	598	656	711	765
0.60	2.25	261	347	428	503	574	642	707	770	830	889
0.80	2.41	329	421	505	584	659	730	798	864	927	989
1.00	2.61	388	480	564	642	716	787	854	919	982	1043
1.20	2.90	373	451	523	590	653	713	770	826	879	931

Remaining Requirement for Escape Protein, g/d

Gain kg/day	ME mcal/kg	150	200	250	300	350	400	450	500	550	600
Medium frame steer											
0.20	1.94	261	258	275	251	249	246	243	241	239	
0.40	2.08	320	299	339	262	245	229	213	199	184	
0.60	2.22	361	323	412	255	224	194	165	137	111	
0.80	2.36	389	335	483	238	193	151	110	71	33	
1.00	2.53	417	346	548	220	162	107	54	3	0	
1.20	2.74	458	375	566	227	158	92	29	0	0	
Large frame steer											
0.20	1.88	312	311	312	312	312	312	312	312	312	312
0.40	2.00	373	388	359	345	332	320	308	297	286	276
0.60	2.11	413	443	385	359	333	309	286	264	242	221
0.80	2.23	442	488	399	358	319	282	247	212	179	147
1.00	2.35	459	521	402	348	296	248	201	155	111	69
1.20	2.49	474	551	402	335	271	210	152	95	40	0
1.40	2.65	496	586	412	334	260	188	120	54	0	0
1.60	2.87	651	553	463	377	296	218	143	70	0	0
Medium frame bull											
0.20	1.92	265	264	263	262	261	260	260	259	258	258
0.40	2.05	337	321	305	290	276	263	250	238	227	215
0.60	2.17	390	358	329	301	275	250	226	203	181	159
0.80	2.30	434	387	343	302	263	226	190	155	122	90
1.00	2.44	473	410	353	299	247	199	151	106	62	20
1.20	2.60	513	438	368	302	240	181	124	68	15	0
1.40	2.82	577	492	413	339	268	201	135	72	11	0
Large frame bull											
0.20	1.89	296	297	297	298	298	299	299	300	300	301
0.40	2.00	378	365	353	341	330	320	310	301	292	283
0.60	2.10	441	415	391	368	347	326	306	287	269	251
0.80	2.21	498	458	421	386	352	321	290	260	232	204
1.00	2.32	546	492	442	395	351	308	267	228	190	153
1.20	2.45	593	525	462	402	346	292	241	191	143	96
1.40	2.58	636	557	483	414	348	285	225	167	110	55
1.60	2.76	699	610	527	449	375	304	236	170	106	43
1.80	3.07	819	730	647	569	494	421	351	283	217	152
Medium frame heifer											
0.20	2.01	205	198	191	185	180	174	169	165	160	
0.40	2.20	240	212	186	161	138	116	95	75	56	
0.60	2.39	256	208	164	123	84	47	11	0	0	
0.80	2.62	273	208	148	91	37	0	1	0	0	
1.00	3.15	382	313	247	185	126	68	13	0	0	
Large frame heifer											
0.20	1.94	255	251	248	245	242	239	237	235	232	230
0.40	2.09	306	285	264	245	227	210	194	178	163	148
0.60	2.25	338	297	258	222	188	155	124	94	65	7
0.80	2.41	355	296	241	190	141	94	50	6	0	0
1.00	2.61	375	299	228	162	99	39	0	0	0	0
1.20	2.90	432	346	266	191	119	50	0	0	0	0

of body weight of which 1.43 and .97 percent of body weight are concentrate (CI) and forage (FI) intakes, respectively. From these values, efficiency of microbial protein (BCP) synthesis is 117 g BCP per kg TDN based on the equation: $6.25 * (8.63 + 14.6FI - 5.18FI * FI + 0.59 * CI)$. From the 7.2 kg of feed at 75.9 percent TDN, TDN intake is 5.47 kg. Total BCP synthesis in the rumen per day therefore is 641 g as shown in Table 12.

Of this, 20 percent is nonprotein material, making true microbial protein (TBP) yield 512 g. At 80 percent digestibility, only 410 g become available as digestible bacterial protein (DBP). To determine how much dietary protein is needed to supplement DBP to meet the protein requirement for this steer, protein requirements must be estimated next.

Metabolic fecal protein (FPN) loss was calculated as

0.09 indigestible dry matter. Indigestible dry matter in this example is 1.73 kg (7.2 kg feed minus 5.47 kg TDN) so that FPN equals 156 g. Endogenous urinary protein loss (UPN) at 2.75 $W^{0.5}$ for this steer would be 48 g and surface losses (SPN) at 0.2 $W^{0.6}$ would be 6 g. Conversion of available amino acids to FPN was assumed to be 100 percent, while conversion to UPN and SPN was estimated to be 0.67. Hence, the inevitable protein loss would total 80 g. Amount of protein deposited in tissue for a 300-kg steer gaining 1.2 kg per day calculated from the NRC (1984) equation (268 − (29.4 * energy content of gain)) * daily gain, with 4.91 Mcal NE_g deposited with 1.2 kg or 4.09 Mcal per kg gain, would be 178 g. Protein deposited in tissue calculated from the ARC (1980) equation also is predicted to be 178 g. [Greater deviation is found with heavier cattle for which the NRC (1984) equation predicts considerably lower rates of protein deposition.] Assuming an efficiency of con-

APPENDIX TABLE 8 Requirements for Total and Escape Protein for Beef Cattle of Various Types, Weights, and Rates of Gain

Dietary Protein Requirement, % of DM

Daily gain kg/day	\<150\>	\<200\>	\<250\>	\<300\>	\<350\>	\<400\>	\<450\>	\<500\>	\<550\>	\<600\>
Medium frame steer										
0.20	9.61	9.09	9.05	8.47	8.29	8.12	8.00	7.89	7.79	
0.40	11.41	10.32	10.49	9.13	8.76	8.45	8.20	8.01	7.84	
0.60	13.12	11.52	12.24	9.76	9.21	8.77	8.41	8.11	7.87	
0.80	14.78	12.65	14.01	10.33	9.60	9.03	8.57	8.19	7.87	
1.00	16.55	13.84	15.77	10.89	9.97	9.25	8.67	8.19	8.14	
1.20	18.29	14.97	16.55	11.34	10.20	9.31	8.59	8.27	8.21	
Large frame steer										
0.20	9.39	8.91	8.64	8.42	8.25	8.11	8.01	7.91	7.83	7.76
0.40	10.85	10.44	9.57	9.12	8.78	8.53	8.29	8.13	7.97	7.84
0.60	12.19	11.84	10.42	9.78	9.28	8.88	8.57	8.32	8.09	7.90
0.80	13.65	13.30	11.36	10.46	9.80	9.28	8.86	8.50	8.21	7.97
1.00	15.03	14.73	12.22	11.13	10.29	9.64	9.12	8.68	8.32	8.01
1.20	16.56	16.21	13.14	11.79	10.79	10.00	9.36	8.83	8.38	8.08
1.40	18.09	17.61	13.98	12.39	11.20	10.25	9.51	8.88	8.42	8.38
1.60	21.67	17.46	14.77	12.86	11.45	10.32	9.43	8.68	8.07	7.90
Medium frame bull										
0.20	9.67	9.16	8.82	8.58	8.39	8.23	8.11	8.01	7.90	7.83
0.40	11.63	10.58	9.87	9.36	8.98	8.69	8.44	8.23	8.07	7.91
0.60	13.40	11.82	10.79	10.05	9.52	9.08	8.73	8.44	8.20	7.98
0.80	15.21	13.11	11.72	10.77	10.04	9.47	9.01	8.62	8.30	8.02
1.00	17.05	14.38	12.68	11.46	10.54	9.83	9.25	8.77	8.37	8.03
1.20	18.87	15.63	13.55	12.08	10.97	10.12	9.42	8.84	8.35	8.20
1.40	20.84	16.90	14.39	12.62	11.28	10.24	9.40	8.70	8.12	7.97
Large frame bull										
0.20	9.72	9.22	8.88	8.65	8.45	8.31	8.18	8.08	7.99	7.91
0.40	11.67	10.63	9.95	9.46	9.09	8.82	8.58	8.39	8.22	8.09
0.60	13.39	11.87	10.91	10.21	9.69	9.27	8.93	8.66	8.43	8.23
0.80	15.21	13.20	11.91	10.99	10.29	9.76	9.31	8.94	8.64	8.37
1.00	17.01	14.51	12.87	11.73	10.88	10.19	9.66	9.21	8.83	8.49
1.20	18.95	15.91	13.93	12.53	11.47	10.67	10.00	9.45	9.00	8.60
1.40	20.72	17.17	14.85	13.21	11.99	11.02	10.27	9.62	9.09	8.62
1.60	22.82	18.57	15.84	13.92	12.48	11.35	10.44	9.69	9.06	8.51
1.80	25.12	19.99	16.71	14.40	12.66	11.30	10.21	9.31	8.56	7.91
Medium frame heifer										
0.20	9.68	9.09	8.68	8.40	8.20	8.03	7.88	7.77	7.66	
0.40	11.43	10.24	9.46	8.92	8.49	8.16	7.88	7.66	7.48	
0.60	13.00	11.24	10.12	9.31	8.70	8.22	7.83	7.71	7.70	
0.80	14.55	12.19	10.67	9.57	8.75	8.26	8.21	8.14	8.09	
1.00	15.99	12.65	10.48	8.95	7.81	6.92	6.21	6.03	5.99	
Large frame heifer										
0.20	9.47	8.96	8.63	8.39	8.21	8.04	7.86	7.83	7.73	7.65
0.40	11.15	10.11	9.44	8.96	8.59	8.30	8.07	7.88	7.71	7.56
0.60	12.88	11.28	10.26	9.52	8.98	8.53	8.19	7.89	7.64	7.21
0.80	14.47	12.34	10.97	10.01	9.28	8.71	8.24	7.86	7.80	7.79
1.00	16.15	13.45	11.67	10.43	9.50	8.77	8.34	8.29	8.25	8.21
1.20	17.78	14.33	12.12	10.56	9.39	8.47	7.88	7.81	7.73	7.68

Dietary Escape Protein Requirement, % of Protein

Daily gain kg/day	\<150\>	\<200\>	\<250\>	\<300\>	\<350\>	\<400\>	\<450\>	\<500\>	\<550\>	\<600\>
Medium frame steer										
0.20	71	60	54	46	42	38	35	33	30	
0.40	69	58	54	42	37	32	28	25	22	
0.60	65	54	54	37	31	25	20	16	13	
0.80	61	49	54	32	25	19	13	8	4	
1.00	58	46	54	28	20	13	6	0	0	
1.20	59	47	54	28	19	11	3	0	0	
Large frame steer										
0.20	84	71	62	56	51	47	43	40	38	36
0.40	82	71	61	53	48	43	39	35	32	30
0.60	77	69	57	50	43	38	33	29	26	23
0.80	72	65	53	45	38	32	27	22	18	14
1.00	66	62	48	40	33	27	21	16	11	7
1.20	61	59	45	36	28	22	15	9	4	0
1.40	59	58	43	34	26	19	12	5	0	0
1.60	67	57	48	39	31	23	15	7	0	0
Medium frame bull										
0.20	73	62	54	48	44	40	37	35	33	31
0.40	72	61	53	46	41	36	33	29	27	24
0.60	70	59	50	43	37	32	27	23	20	17
0.80	67	56	47	39	32	27	22	17	13	9
1.00	64	53	44	36	29	22	16	11	6	2
1.20	63	52	43	34	27	20	13	7	2	0
1.40	66	56	46	38	30	22	15	8	1	0
Large frame bull										
0.20	79	68	59	53	49	45	42	39	37	35
0.40	80	68	60	53	47	43	39	36	33	30
0.60	78	67	58	51	45	40	36	32	29	26
0.80	75	64	55	48	42	36	31	27	23	20
1.00	72	61	53	45	38	33	27	23	18	14
1.20	69	59	50	42	35	29	23	18	13	9
1.40	68	58	49	41	34	27	21	16	10	5
1.60	69	60	51	43	36	29	22	16	10	4
1.80	80	72	65	58	51	44	37	30	24	17
Medium frame heifer										
0.20	58	48	41	36	32	29	26	24	22	
0.40	54	43	35	28	22	18	14	10	7	
0.60	49	37	28	20	13	7	2	0	0	
0.80	47	34	24	14	6	0	0	0	0	
1.00	68	57	46	35	24	13	3	0	0	
Large frame heifer										
0.20	71	59	52	46	41	37	34	32	30	28
0.40	68	56	47	40	34	30	26	23	20	17
0.60	62	50	40	33	26	21	16	11	8	1
0.80	57	45	35	26	19	12	6	1	0	0
1.00	53	41	31	21	13	5	0	0	0	0
1.20	59	47	36	26	16	7	0	0	0	0

version of retained metabolizable protein (RPM) to retained net protein (RNP) of 0.50, 355 g of absorbed protein would be needed here. The requirements for maintenance (80), fecal loss (156), and gain (355) total 591 g and must be supplied by intestinal digestion of either dietary or microbial protein.

With a total need of 591 g and a supply of digestible true protein from ruminal microbes (DBP) of 410 g, the remaining deficit, which must be supplied as digestible dietary protein that has escaped ruminal fermentation (DUP), is 181 g. Assuming that such escape protein has a digestibility of 0.80, the supply of escape protein (IUP) must be 227 g as shown in Appendix Table 7. In addition to this dietary protein need, a source of protein (or NPN) is needed for microbial protein synthesis. To syn-

thesize 641 g BCP, assuming a capture efficiency of 0.90, primarily due to ammonia loss with liquid flowing from the rumen, 712 g of ruminally digested protein (RAP) is needed. Besides the supply from the diet, some protein is recycled to the rumen. Recycling by the regression equation suggests that 33 percent of IP would be available by recylcing leading to a very low (9.56 percent) protein requirement. Using the recycling value of 15 percent of IP, as used for dairy cattle, the total dietary protein need is (RAP + UIP)/1.15 or 817 g or as a percent of diet, 11.34 as shown in Appendix Table 8. Of the dietary protein, 227 g must escape ruminal fermentation. This is 28 percent of the dietary protein, also as shown in Appendix Table 8.

Description of Appendix Tables 9 and 10

Appendix Table 9 presents a rigorous mathematical statement of the factors adopted for transforming feed protein into net protein for dairy cattle. A FORTRAN IV source program is used that includes the necessary input and output statements. Nearly every statement is documented with a comment statement. The energy requirement, net protein requirement for growth, and dry matter intake for growth are from the NRC (1978) requirements. The metabolic fecal protein equivalent, intake protein and metabolic transformations are those developed in this report. Examples are given for the 300-kg Holstein heifer gaining .7 kg/day and the 600-kg cow producing 30 kg of 3.5 percent milk/day.

Appendix Table 10 presents a rigorous mathematical statement of the factors adopted for transforming feed protein into net protein for beef cattle. The rigorous mathematical statement is a working FORTRAN IV source program that includes the necessary input and output statements. Nearly every statement is documented with a comment statement. The energy requirement, net protein requirement, and dry matter intake are from the NRC (1984) requirements. The metabolic fecal protein equivalent, intake protein and metabolic transformations are those developed in this report. Examples are given for the 300-kg medium-frame steer calf gaining 1.2 kg/day, the 500-kg cow that is 225 days pregnant and expected to produce a calf weighing 36 kg, and the 500-kg cow producing 10 kg of 4.0 percent milk/day.

APPENDIX TABLE 9   Factors Adopted for Transforming Feed Protein into Net Protein for Dairy Cattle

```
      DATE
13 Dec 84 14:29:44 Thursday
OK, SLIST NRCD78.FTN
C FORTRAN IV PROGRAM FOR PROTEIN REQUIREMENT BASED ON NRC N SUBCOMMITTEE
C AS RUN ON A PRIME 750 SYSTEM. OTHER SYSTEMS WILL REQUIRE INPUT-OUTOUT
C CHANGES.
C ENERGY REQUIREMENT FROM DAIRY NRC (1978).
C NET PROTEIN REQUIREMENT PARTLY FROM DAIRY NRC (1978).
C DRY MATTER INTAKE FOR GROWTH FROM DAIRY NRC (1978).
C METABOLIC FECAL PROTEIN, INTAKE PROTEIN AND PROTEIN METABOLISM BASED
C ON NITROGEN NRC.
C /*ARE COMMENT STATEMENTS ALLOWED ON-LINE IN PRIME SYSTEMS.
$INSERT SYSCOM¶KEYS.F
C VARIABLES DEFINED TO BREAK DEFAULT TYPES.
      REAL IBP,IDM,IIP,IIPIP,INP,IOM,IP,IPDM,IUP,JOUR,LPA,LPI,LPN,
     2LNCONV,LPNIP,LPNKG,LPNLPA,LNEMKG,LUTERU,ME,MEADM,MEDM,MILKKG,MPA,
     3MPN,MPNMPA,NCP,NCPBCP,NEG,NEGADM,NEGAIN,NEL,NEM,NEMADM
C
C DATA INPUT OR DEFINITION OF ANIMAL
C
      WRITE(1,5)
    5 FORMAT(1X,'FOR GROWING ANIMALS THIS PROGRAM IS RESTRICTED TO:')
      WRITE(1,6)
    6 FORMAT(3X,'WEIGHTS OF 100,150,200,250,300,350,400,450,500,550 KG')
      WRITE(1,7)
    7 FORMAT(3X,'AND A GAIN OF 0.70 KG/DAY.')
      WRITE(1,10)
   10 FORMAT(1X,'ENTER BODY WEIGHT IN KG AS XXX. ')
      READ(1,11)BW
   11 FORMAT(F5.0)
      WRITE(1,12)
   12 FORMAT(1X,'ENTER MILK PRODUCTION IN KG AS XX. ')
      READ(1,11)MILKKG
      WRITE(1,13)
   13 FORMAT(1X,'ENTER MILK FAT TEST % AS X.XX ')
      READ(1,14)PFAT
   14 FORMAT(F5.2)
      WRITE(1,15)
   15 FORMAT(1X,'ENTER DAYS PREGNANT AS XXX. ')
      READ(1,11)DAYS
      WRITE(1,16)
   16 FORMAT(1X,'ENTER DAILY GAIN OR LOSS (-) IN KG AS X.XX ')
      READ(1,14)DBW
      WRITE(1,17)
   17 FORMAT(1X,'ENTER DAILY WEIGHT GAIN IN KG AS X.XX ')
      READ(1,14)GAIN
C
C DRY MATTER AND ENERGY RELATIONSHIPS
C
      BTDNM=.0352*BW**.75           /*BASELINE TDN FOR MAINTENANCE, KG
      IF(GAIN.GT.0.0)BTDNM=0.0
      FCMKG=.4+.15*PFAT
      FCM=MILKKG*FCMKG
      BTDNL=.326*FCM                /*BASELINE TDN FOR LACTATION, KG
      BTDNP=.0106*BW**.75           /*BASELINE TDN FOR PREGNANCY, KG
      IF(DAYS.LE.210)BTDNP=0.0
      IF(DBW.LT.0)BTDNDM=2.17       /*BASELINE TDN PER KG LOSS IN LACTATION
      IF(DBW.GT.0)BTDNDM=2.26       /*BASELINE TDN PER KG GAIN IN LACTATION
      BTDND=BTDNDM*DBW              /*BASELINE TDN FOR - OR + IN LACTATION, KG
      BTDNR=0.0
      IF(GAIN.EQ.0.0)GO TO 21
C ITERATIVE LOOP FOR SOLVING ME FOR GROWTH, MCAL ME/KG AIR DM
      N=0                           /*SETS ITERATION COUNTER TO 0
      MEADM=1.8                     /*SETS INITIAL ME AT 2.0 MCAL.KG AIR DM
      IF(BW.EQ.100.)DM=2.80         /*DM. KG/DAY
```

APPENDIX TABLE 9   (*Continued*)

```
              IF(BW.EQ.150.)DM=4.00
              IF(BW.EQ.200.)DM=5.20
              IF(BW.EQ.250.)DM=6.30
              IF(BW.EQ.300.)DM=7.20
              IF(BW.EQ.350.)DM=8.00
              IF(BW.EQ.400.)DM=8.60
              IF(BW.EQ.450..)DM=9.10
              IF(BW.EQ.500.)DM=9.50
              IF(BW.EQ.550.)DM=9.80
              NEG=NEGAIN*GAIN               /*REQUIRED NEG, MCAL
              AIRDM=DM/.9                   /*CALCULATES AIR DM FROM DM
              NEM=.077*BW**.75              /*REQUIRED NEM, MCAL
              IF(BW.EQ.100.)NEGAIN=2.10      /*NEGAIN, MCAL/KG GAIN
              IF(BW.EQ.150.)NEGAIN=2.40      /*READ FROM FIGURE 1 (NRC, 1978)
              IF(BW.EQ.200.)NEGAIN=2.80
              IF(BW.EQ.250.)NEGAIN=3.10
              IF(BW.EQ.300.)NEGAIN=3.40
              IF(BW.EQ.350.)NEGAIN=3.60
              IF(BW.EQ.400.)NEGAIN=3.80
              IF(BW.EQ.450.)NEGAIN=4.00
              IF(BW.EQ.500.)NEGAIN=4.20
              IF(BW.EQ.550.)NEGAIN=4.40
              IF(BW.EQ.600.)NEGAIN=4.50
              NEG=NEGAIN*GAIN               /*REQUIRED NEG, MCAL
           30 CONTINUE
      C CALCULATIONS BASED ON EQUATION DERIVED FROM LOFGREEN AND GARRET (1968)
      C CONVERT Y AXIS OF EQUATION IN FIGURE 5 TO NEM, MCAL/KG AIR DM
      C LOG10 77 - LOG10 Y = 2.3030 - .2455 X
      C LOG10 Y =LOG10 77 - 2.3030 + .2455 X
      C LOG10 Y = -.4165 + .2455 X
      C CONVERT LOG10 Y TO LOGE Y
      C LOGE Y = LOGE 10 * (-.4165 + .2455 X)
      C LOGE Y = 2.303 * (-.4165 + .2455 X)
      C LOGE Y = .5653 X - .9590
              LNEMKG=.5653*MEADM-.9590      /*CALCULATES LOGE OF NEM, MCAL/KG AIR DM
              NEMADM=EXP(LNEMKG)            /*CALCULATES NEM, MCAL/KG AIR DM
      C CONVERT X AXIS OF EQUATION IN FIGURE 6 TO NEM, MCAL/KG AIR DM
      C Y = 2.29 - 77 * (.0254/X)
      C Y = 2.29 - 1.955/X
              NEGADM=2.29-1.9558/NEMADM     /*CALCULATES NEG, MCAL/KG AIR DM
              ADMM=NEM/NEMADM               /*AIR DM FOR MAINTENANCE, KG
              ADMG=NEG/NEGADM               /*AIR DM FOR GROWTH, KG
              ADMT=ADMM+ADMG                /*TOTAL AIR DM REQUIRED, KG
              ME=MEADM*ADMT                 /* CALCULATES ME, MCAL
              N=N+1                         /*COUNTS ITERATION CYCLE
              IF(N.GT.25)GO TO 60           /*SETS ITERATION LIMIT
              IF(ABS(AIRDM-ADMT).LE. .02)GO TO 60 /*DECISION TO END ITERATION
              IF(AIRDM.LT.ADMT)GO TO 40     /*DECISION TO INCREASE MEADM
              IF(AIRDM.GT.ADMT)GO TO 50     /*DECISION TO DECREASE MEADM
           40 CONTINUE
              MEADM=MEADM+.1                /*INCREASE MEADM BY +.1
              GO TO 30
           50 CONTINUE
              MEADM=MEADM-.01               /*DECREASE MEADM BY -.01
              GO TO 30
      C END OF ITERATIVE LOOP
           60 CONTINUE
              MEDM=MEADM*.9                 /*ME, MCAL/KG DM
              MEDM=MEDM/(1-.08)             /*ADJUSTS FOR DIGESTIBILITY DEPRESSION
              DEDM=MEDM/1.01+.45            /*DE, MCAL/KG DM
              DE=DM*DEDM                    /*DE, MCAL
              BTDNR=DE/4.409                /*BASELINE TDN FOR RETENTION, KG
           21 CONTINUE
              BTDN=BTDNM+BTDNL+BTDNP+BTDND+BTDNR    /*BASELINE TDN (TOTAL), KG
              DEPRES=.08
              ATDN=BTDN*(1-DEPRES)          /*ADJUSTED TDN. KG
```

APPENDIX TABLE 9   (*Continued*)

```
      IF(GAIN.GT.0.0) GO TO 70
      DM=BTDN/.75
   70 CONTINUE
      IDM=DM-ATDN
C
C STATED PROTEIN FACTORS WITH PROPORTIONAL UNITS
C
      BTPBCP=.80     /*BACTERIAL TRUE PROTEIN/BACTERIAL CRUDE PROTEIN
      DBPBTP=.80     /*DIGESTIBLE BACTERIAL PROTEIN/BACTERIAL TRUE PROTEIN
      DUPUIP=.80     /*DIGESTIBLE UNDEGRADED PROTEIN/UNDEGRADED INTAKE PROTEIN
      MPNMPA=.67     /*MAINTENANCE PROTEIN NET/MAINTENANCE PROTEIN ABSORBED
      YPNYPA=.50     /*CONCEPTUS PROTEIN NET/CONCEPTUS PROTEIN ABSORBED
      FPAIDM=.090    /*(METABOLIC) FECAL PROTEIN ABSORBED/INDIGESTIBLE DRY MATTER
      DNPNCP=1.00    /*DIGESTIBLE NUCLEIC PROTEIN/NUCLEIC CRUDE PROTEIN
      BCPRAP=.90     /*BACTERIAL CRUDE PROTEIN/RUMEN AVAILABLE PROTEIN
      RIPIP=.15      /*RUMEN INFLUX PROTEIN/INTAKE PROTEIN
      LPNLPA=.65     /*LACTATION PROTEIN NET/LACTATION PROTEIN ABSORBED
      RPNRPA=.50     /*RETAINED PROTEIN NET/RETAINED PROTEIN ABSORBED
C
C CALCULATION OF PROTEIN REQUIREMENTS
C
      SPN= .2 *BW**.6              /*SCURF PROTEIN NET, G
      UPN=2.75*BW**.5             /*ENDOGENOUS URINARY PROTEIN NET, G
      SPA=SPN/MPNMPA              /*SCURF PROTEIN ABSORBED, G
      UPA=UPN/MPNMPA             /*ENDOGENOUS URINARY PROTEIN ABSORBED, G
      MPA=SPA+UPA                /*MAINTENANCE  PROTEIN ABSORBED, G
      FPA=IDM*FPAIDM*1000        /*METABOLIC FECAL PROTEIN ABSORBED, G
      LPNKG=(1.9+.4*PFAT)/100
      LPN=MILKKG*LPNKG*1000      /*LACTATION PROTEIN NET, G
      LPA=LPN/LPNLPA             /*LACTATION PROTEIN ABSORBED, G
      YPN=1.136*BW**.7           /*CONCEPTUS PROTEIN NET, G
      IF(DAYS.LE.210)YPN=0.0
      YPA=YPN/YPNYPA             /*CONCEPTUS PROTEIN ABSORBED, G
      IF(BW.EQ.100.)RPNLWG=.175   /*RPNLWG, PROPORTIONAL
      IF(BW.EQ.150.)RPNLWG=.168   /*READ FROM FIGURE 1 (NRC, 1978)
      IF(BW.EQ.200.)RPNLWG=.166
      IF(BW.EQ.250.)RPNLWG=.164
      IF(BW.EQ.300.)RPNLWG=.162
      IF(BW.EQ.350.)RPNLWG=.161
      IF(BW.EQ.400.)RPNLWG=.160
      IF(BW.GE.450.)RPNLWG=.159
      RPN=RPNLWG*GAIN*1000       /*RETAINED PROTEIN NET, G
      RPA=RPN/RPNRPA             /*RETAINED PROTEIN ABSORBED, G
      DPA=160*DBW                /*DIFFERENCE PROTEIN FROM DBW IN LACTATION, G
      AP=MPA+LPA+FPA+YPA+RPA+DPA  /*ABSORBED PROTEIN (TOTAL), G
C FLOW OF PROTEIN
      BCP=6.25*(-31.86+26.12*BTDN) /*BACTERIAL CRUDE PROTEIN, G
      RAP=BCP/BCPRAP             /*RUMEN AVAILABLE PROTEIN, G
      BTP=BCP*BTPBCP             /*BACTERIAL TRUE PROTEIN, G
      DBP=BTP*DBPBTP             /*DIGESTIBLE BACTERIAL PROTEIN, G
      IBP=BTP-DBP                /*INDIGESTIBLE BACTERIAL PROTEIN, G
      NCP=BCP-BTP                /*NUCLEIC CRUDE PROTEIN, G
      DNP=NCP*DNPNCP             /*DIGESTIBLE NUCLEIC PROTEIN, G
      INP=NCP-DNP                /*INDIGESTIBLE NUCLEIC PROTEIN, G
      DUP=AP-DBP                 /*DIGESTIBLE UNDEGRADED PROTEIN, G
      UIP=DUP/DUPUIP             /*UNDEGRADED INTAKE PROTEIN, G
      IUP=UIP-DUP                /*INDIGESTIBLE UNDEGRADED PROTEIN, G
      REP=RAP*(1-BCPRAP)         /*RUMEN EFFLUX PROTEIN, G
      LPI=LPA-LPN                /*LACTATION PROTEIN INCREMENT, G
      YPI=YPA-YPN                /*CONCEPTUS PROTEIN INCREMENT, G
      RPI=RPA-RPN                /*RETAINED PROTEIN INCREMENT, G
      IP=(RAP+UIP)/(1+RIPIP)     /*INTAKE PROTEIN, G
      IPDM=IP/(DM*1000)          /*INTAKE PROTEIN/DRY MATTER, PROPORTIONAL
      RIP=RIPIP*IP               /*RUMEN INFLUX PROTEIN, G
      DIP=RAP-RIP                /*DEGRADABLE INTAKE PROTEIN, G
      UIPIP=UIP/IP               /*UNDEGRADED INTAKE PROTEIN/INTAKE PROTEIN. G
```

APPENDIX TABLE 9   (*Continued*)

```
C OUTPUT OF PROTEIN
      FP=IBP+INP+IUP+FPA              /*FECAL PROTEIN, G
      UP=REP+DNP+MPA+LPI+YPI+RPI-RIP-SPN    /*URINARY PROTEIN, G
      LPNIP=LPN/IP              /*LACTATION PROTEIN NET/INTAKE PROTEIN, PROPORTIONAL
      YPNIP=YPN/IP              /*CONCEPTUS PROTEIN NET/INTAKE PROTEIN, PROPORTIONAL
      RPNIP=RPN/IP              /*RETAINED PROTEIN NET/INTAKE PROTEIN, PROPORTIONAL
      DPAIP=DPA/IP              /*DIFFERENCE PROTEIN ABS/INTAKE PROTEIN, PROPORTION
      SPNIP=SPN/IP              /*SCURF PROTEIN NET/INTAKE PROTEIN, PROPORTIONAL
      UPIP=UP/IP               /*URINARY PROTEIN/INTAKE PROTEIN, PROPORTIONAL
      FPIP=FP/IP               /*FECAL PROTEIN/INTAKE PROTEIN, PROPORTIONAL
C PREDICTED PROTEIN AT DUODENUM
      STP=BTP+UIP                   /*SMALL (INTESTINE) TRUE PROTEIN, G
      SCP=BCP+UIP                   /*SMALL (INTESTINE) CRUDE PROTEIN, G
C EXPECTED PROTEIN AT DUODENUM
      TAMM=((32.3*DOM)-8.63)*6.25
      ROHR=((31.42*DOM)-40.56)*6.25
      JOUR=((22.62*DOM)+(.687*UIP/6.25)+4.3)*6.25
      VERI=((23.85*DOM)+(.600*UIP/6.25)+8.6)*6.25
C
C DATA OUTPUT OR PRINTING OF RESULTS
C
      WRITE(1,90)
   90 FORMAT(/)
      WRITE(1,100)IPDM
  100 FORMAT('INTAKE PROTEIN/DRY MATTER IS                  ',F5.4,)
      WRITE(1,110)UIPIP
  110 FORMAT('UNDEGRADED INTAKE PROTEIN/INTAKE PROTEIN IS       ',F5.3,)
      WRITE(1,120)BTDN
  120 FORMAT('BASELINE TDN INTAKE IS                   ',F5.1,' KG')
      WRITE(1,130)DM
  130 FORMAT('DRY MATTER INTAKE IS                    ',F5.1,' KG')
      IF(GAIN.EQ.0.0) GO TO 131
      WRITE(1,140)MEADM
  140 FORMAT('MCAL ME/KG AIR DM IS                    ',F5.2,)
      WRITE(1,150)NEMADM
  150 FORMAT('MCAL NEM/KG AIR DM IS                    ',F5.2,)
      WRITE(1,160)NEGADM
  160 FORMAT('MCAL NEG/KG AIR DM IS                    ',F5.2,)
      WRITE(1,170)ME
  170 FORMAT('MCAL ME NEEDED IS                     ',F5.2,)
      WRITE(1,180)NEM
  180 FORMAT('MCAL NEM NEEDED IS                    ',F5.2,)
      WRITE(1,190)NEG
  190 FORMAT('MCAL NEG NEEDED IS                    ',F5.2,)
  131 CONTINUE
      WRITE(1,90)
      WRITE(1,200)SPA
  200 FORMAT('SURFACE PROTEIN IN ABSORBED UNITS IS          ',F5.0,' G')
      WRITE(1,210)UPA
  210 FORMAT('URINARY PROTEIN IN ABSORBED UNITS IS          ',F5.0,' G')
      WRITE(1,220)MPA
  220 FORMAT('MAINTENANCE PROTEIN IN ABSORBED UNITS IS       ',F5.0,' G')
      WRITE(1,230)FPA
  230 FORMAT('MET. FECAL PROTEIN IN ABSORBED UNITS IS       ',F5.0,' G')
      WRITE(1,240)LPA
  240 FORMAT('LACTATION PROTEIN IN ABSORBED UNITS IS        ',F5.0,' G')
      WRITE(1,250)YPA
  250 FORMAT('CONCEPTUS PROTEIN IN ABSORBED UNITS IS        ',F5.0,' G')
      WRITE(1,260)DPA
  260 FORMAT('DIFFERENCE PROTEIN IN ABSORBED UNITS IS       ',F5.0,' G')
      WRITE(1,270)RPA
  270 FORMAT('RETAINED PROTEIN IN ABSORBED UNITS IS         ',F5.0,' G')
      WRITE(1,280)AP
  280 FORMAT('REQUIRED PROTEIN IN ABSORBED UNITS IS         ',F5.0,' G')
      WRITE(1,90)
      WRITE(1.300)BCP
```

APPENDIX TABLE 9   *(Continued)*

```
300 FORMAT('BACTERIAL CRUDE PROTEIN IS                    ',F5.0,' G')
    WRITE(1,310)NCP
310 FORMAT('NUCLEIC CRUDE PROTEIN IS                      ',F5.0,' G')
    WRITE(1,320)BTP
320 FORMAT('BACTERIAL TRUE PROTEIN IS                     ',F5.0,' G')
    WRITE(1,330)DBP
330 FORMAT('DIGESTIBLE BACTERIAL PROTEIN IS               ',F5.0,' G')
    WRITE(1,340)DUP
340 FORMAT('DIGESTIBLE UNDEGRADED PROTEIN IS              ',F5.0,' G')
    WRITE(1,350)UIP
350 FORMAT('UNDEGRADED INTAKE PROTEIN IS                  ',F5.0,' G')
    WRITE(1,360)DIP
360 FORMAT('DEGRADED INTAKE PROTEIN IS                    ',F5.0,' G')
    WRITE(1,370)IP
370 FORMAT('INTAKE PROTEIN IS                             '
   2,F5.0,' G')
    WRITE(1,90)
    WRITE(1,380)SPN
380 FORMAT('SCURF PROTEIN IS                              '
   2,F5.0,' G')
    WRITE(1,400)RIP
400 FORMAT('RUMEN INFLUX PROTEIN IS                       ',F5.0,' G')
    WRITE(1,410)REP
410 FORMAT('RUMEN EFFLUX PROTEIN IS                       ',F5.0,' G')
    WRITE(1,420)DNP
420 FORMAT('DIGESTIBLE NUCLEIC PROTEIN IS                 ',F5.0,' G')
    WRITE(1,430)MPA
430 FORMAT('MAINTENANCE PROTEIN IN ABSORBED UNITS IS      ',F5.0,' G')
    WRITE(1,440)LPI
440 FORMAT('LACTATION PROTEIN INCREMENT IS                ',F5.0,' G')
    WRITE(1,450)YPI
450 FORMAT('PREGNANCY PROTEIN INCREMENT IS                ',F5.0,' G')
    WRITE(1,460)RPI
460 FORMAT('RETAINED PROTEIN INCREMENT IS                 ',F5.0,' G')
    WRITE(1,470)UP
470 FORMAT('URINARY PROTEIN IS                            '
   2,F5.0,' G')
    WRITE(1,90)
    WRITE(1,500)IBP
500 FORMAT('INDIGESTIBLE BACTERIAL PROTEIN IS             ',F5.0,' G')
    WRITE(1,510)INP
510 FORMAT('INDIGESTIBLE NUCLEIC PROTEIN IS               ',F5.0,' G')
    WRITE(1,520)IUP
520 FORMAT('INDIGESTIBLE UNDEGRADED PROTEIN IS            ',F5.0,' G')
    WRITE(1,530)FPA
530 FORMAT('METABOLIC FECAL PROTEIN IN ABSORBED UNITS IS ',F5.0,' G')
    WRITE(1,540)FP
540 FORMAT('FECAL PROTEIN IS                             '
   2,F5.0,' G')
    WRITE(1,90)
    WRITE(1,600)LPN
600 FORMAT('MILK PROTEIN IN NET UNITS IS                  '
   2,F5.0,' G')
    WRITE(1,610)YPN
610 FORMAT('CONCEPTUS PROTEIN IN NET UNITS IS             '
   2,F5.0,' G')
    WRITE(1,620)DPA
620 FORMAT('DIFFERENCE PROTEIN IN NET UNITS IS            '
   2,F5.0,' G')
    WRITE(1,630)RPN
630 FORMAT('RETAINED PROTEIN IN ABSORBED UNITS IS         '
   2,F5.0,' G')
    WRITE(1,90)
    WRITE(1,700)LPNIP
700 FORMAT('MILK PROTEIN IN NET UNITS/INTAKE PROTEIN IS   ',F5.3,)
    WRITE(1.710)YPNIP
```

APPENDIX TABLE 9   (*Continued*)

```
710 FORMAT('CONCEPTUS PROTEIN NET/INTAKE PROTEIN IS         ',F5.3,)
    WRITE(1,720)DPAIP
720 FORMAT('DIFFERENCE PROTEIN ABSORBED/INTAKE PROTEIN IS   ',F5.3,)
    WRITE(1,730)RPNIP
730 FORMAT('RETAINED PROTEIN NET/INTAKE PROTEIN IS          ',F5.3,)
    WRITE(1,740)SPNIP
740 FORMAT('SCURF PROTEIN NET/INTAKE PROTEIN IS             ',F5.3,)
    WRITE(1,750)UPIP
750 FORMAT('URINARY PROTEIN/INTAKE PROTEIN IS               ',F5.3,)
    WRITE(1,760)FPIP
760 FORMAT('FECAL PROTEIN/INTAKE PROTEIN IS                 ',F5.3,)
    CALL EXIT
    END
OK,
```

APPENDIX TABLE 9   (*Continued*)

```
   R NRCD78
 FOR GROWING ANIMALS THIS PROGRAM IS RESTRICTED TO:
   WEIGHTS OF 100,150,200,250,300,350,400,450,500,550 KG
   AND A GAIN OF 0.70 KG/DAY.
 ENTER BODY WEIGHT IN KG AS XXX.
300.
 ENTER MILK PRODUCTION IN KG AS XX.
0.
 ENTER MILK FAT TEST % AS X.XX
0.
 ENTER DAYS PREGNANT AS XXX.
0.
 ENTER DAILY GAIN OR LOSS (-) IN KG AS X.XX
0.
 ENTER DAILY WEIGHT GAIN IN KG AS X.XX
.7

 INTAKE PROTEIN/DRY MATTER IS                     .1114
 UNDEGRADED INTAKE PROTEIN/INTAKE PROTEIN IS      0.526
 BASELINE TDN INTAKE IS                           4.0 KG
 DRY MATTER INTAKE IS                             7.2 KG
 MCAL ME/KG AIR DM IS                             2.05
 MCAL NEM/KG AIR DM IS                            1.22
 MCAL NEG/KG AIR DM IS                            0.69
 MCAL ME NEEDED IS                               16.40
 MCAL NEM NEEDED IS                               5.55
 MCAL NEG NEEDED IS                               2.38

 SURFACE PROTEIN IN ABSORBED UNITS IS              9. G
 URINARY PROTEIN IN ABSORBED UNITS IS             71. G
 MAINTENANCE PROTEIN IN ABSORBED UNITS IS         80. G
 MET. FECAL PROTEIN IN ABSORBED UNITS IS         319. G
 LACTATION PROTEIN IN ABSORBED UNITS IS            0. G
 CONCEPTUS PROTEIN IN ABSORBED UNITS IS            0. G
 DIFFERENCE PROTEIN IN ABSORBED UNITS IS           0. G
 RETAINED PROTEIN IN ABSORBED UNITS IS           227. G
 REQUIRED PROTEIN IN ABSORBED UNITS IS           626. G

 BACTERIAL CRUDE PROTEIN IS                      450. G
 NUCLEIC CRUDE PROTEIN IS                         90. G
 BACTERIAL TRUE PROTEIN IS                       360. G
 DIGESTIBLE BACTERIAL PROTEIN IS                 288. G
 DIGESTIBLE UNDEGRADED PROTEIN IS                338. G
 UNDEGRADED INTAKE PROTEIN IS                    422. G
 DEGRADED INTAKE PROTEIN IS                      380. G
 INTAKE PROTEIN IS                                         802. G

 SCURF PROTEIN IS                                            6. G
 RUMEN INFLUX PROTEIN IS                         120. G
 RUMEN EFFLUX PROTEIN IS                          50. G
 DIGESTIBLE NUCLEIC PROTEIN IS                    90. G
 MAINTENANCE PROTEIN IN ABSORBED UNITS IS         80. G
 LACTATION PROTEIN INCREMENT IS                    0. G
 PREGNANCY PROTEIN INCREMENT IS                    0. G
 RETAINED PROTEIN INCREMENT IS                   113. G
 URINARY PROTEIN IS                                        207. G

 INDIGESTIBLE BACTERIAL PROTEIN IS                72. G
 INDIGESTIBLE NUCLEIC PROTEIN IS                   0. G
 INDIGESTIBLE UNDEGRADED PROTEIN IS               84. G
 METABOLIC FECAL PROTEIN IN ABSORBED UNITS IS    319. G
 FECAL PROTEIN IS                                          475. G

 MILK PROTEIN IN NET UNITS IS                                0. G
 CONCEPTUS PROTEIN IN NET UNITS IS                           0. G
```

APPENDIX TABLE 9   (*Continued*)

```
DIFFERENCE PROTEIN IN NET UNITS IS                          0. G
RETAINED PROTEIN IN ABSORBED UNITS IS                     113. G

MILK PROTEIN IN NET UNITS/INTAKE PROTEIN IS      0.000
CONCEPTUS PROTEIN NET/INTAKE PROTEIN IS          0.000
DIFFERENCE PROTEIN ABSORBED/INTAKE PROTEIN IS    0.000
RETAINED PROTEIN NET/INTAKE PROTEIN IS           0.141
SCURF PROTEIN NET/INTAKE PROTEIN IS              0.008
URINARY PROTEIN/INTAKE PROTEIN IS                0.258
FECAL PROTEIN/INTAKE PROTEIN IS                  0.592
OK,
```

APPENDIX TABLE 9   (*Continued*)

```
      R NRCD78
  FOR GROWING ANIMALS THIS PROGRAM IS RESTRICTED TO:
    WEIGHTS OF 100,150,200,250,300,350,400,450,500,550 KG
    AND A GAIN OF 0.70 KG/DAY.
  ENTER BODY WEIGHT IN KG AS XXX.
600.
  ENTER MILK PRODUCTION IN KG AS XX.
30.
  ENTER MILK FAT TEST % AS X.XX
3.5
  ENTER DAYS PREGNANT AS XXX.
0.
  ENTER DAILY GAIN OR LOSS (-) IN KG AS X.XX
0.
  ENTER DAILY WEIGHT GAIN IN KG AS X.XX
0.

INTAKE PROTEIN/DRY MATTER IS                        .1607
UNDEGRADED INTAKE PROTEIN/INTAKE PROTEIN IS        0.381
BASELINE TDN INTAKE IS                            13.3 KG
DRY MATTER INTAKE IS                              17.8 KG

SURFACE PROTEIN IN ABSORBED UNITS IS                14. G
URINARY PROTEIN IN ABSORBED UNITS IS               101. G
MAINTENANCE PROTEIN IN ABSORBED UNITS IS           114. G
MET. FECAL PROTEIN IN ABSORBED UNITS IS            495. G
LACTATION PROTEIN IN ABSORBED UNITS IS            1523. G
CONCEPTUS PROTEIN IN ABSORBED UNITS IS              0. G
DIFFERENCE PROTEIN IN ABSORBED UNITS IS             0. G
RETAINED PROTEIN IN ABSORBED UNITS IS               0. G
REQUIRED PROTEIN IN ABSORBED UNITS IS             2133. G

BACTERIAL CRUDE PROTEIN IS                        1974. G
NUCLEIC CRUDE PROTEIN IS                           395. G
BACTERIAL TRUE PROTEIN IS                         1579. G
DIGESTIBLE BACTERIAL PROTEIN IS                   1264. G
DIGESTIBLE UNDEGRADED PROTEIN IS                   869. G
UNDEGRADED INTAKE PROTEIN IS                      1086. G
DEGRADED INTAKE PROTEIN IS                        1766. G
INTAKE PROTEIN IS                                         2852. G

SCURF PROTEIN IS                                            9. G
RUMEN INFLUX PROTEIN IS                            428. G
RUMEN EFFLUX PROTEIN IS                            219. G
DIGESTIBLE NUCLEIC PROTEIN IS                      395. G
MAINTENANCE PROTEIN IN ABSORBED UNITS IS           114. G
LACTATION PROTEIN INCREMENT IS                     533. G
PREGNANCY PROTEIN INCREMENT IS                      0. G
RETAINED PROTEIN INCREMENT IS                       0. G
URINARY PROTEIN IS                                        825. G

INDIGESTIBLE BACTERIAL PROTEIN IS                  316. G
INDIGESTIBLE NUCLEIC PROTEIN IS                     0. G
INDIGESTIBLE UNDEGRADED PROTEIN IS                 217. G
METABOLIC FECAL PROTEIN IN ABSORBED UNITS IS       495. G
FECAL PROTEIN IS                                         1028. G

MILK PROTEIN IN NET UNITS IS                             990. G
CONCEPTUS PROTEIN IN NET UNITS IS                          0. G
DIFFERENCE PROTEIN IN NET UNITS IS                         0. G
RETAINED PROTEIN IN ABSORBED UNITS IS                      0. G

MILK PROTEIN IN NET UNITS/INTAKE PROTEIN IS       0.347
CONCEPTUS PROTEIN NET/INTAKE PROTEIN IS           0.000
DIFFERENCE PROTEIN ABSORBED/INTAKE PROTEIN IS     0.000
```

APPENDIX TABLE 9   *(Continued)*

```
            RETAINED PROTEIN NET/INTAKE PROTEIN IS      0.000
            SCURF PROTEIN NET/INTAKE PROTEIN IS         0.003
            URINARY PROTEIN/INTAKE PROTEIN IS           0.289
            FECAL PROTEIN/INTAKE PROTEIN IS             0.361
            OK,
```

APPENDIX TABLE 10   Factors Adopted for Transforming Feed Protein into Net Protein for Beef Cattle

```
DATE
13 Dec 84 13:42:00 Thursday
OK, SLIST NRCB84.FTN
C FORTRAN IV PROGRAM FOR PROTEIN REQUIREMENT BASED ON NRC N SUBCOMMITTEE
C AS RUN ON A PRIME 750 SYSTEM. OTHER SYSTEMS WILL REQUIRE INPUT-OUTOUT
C CHANGES.
C ENERGY REQUIREMENT FROM BEEF NRC (1984).
C NET PROTEIN REQUIREMENT FROM BEEF NRC (1984).
C DRY MATTER INTAKE FROM BEEF NRC (1984).
C INTAKE PROTEIN AND PROTEIN METABOLISM BASED ON NITROGEN NRC.
C /* ARE COMMENT STATEMENTS ALLOWED ON-LINE IN PRIME SYSTEMS.
$INSERT SYSCOM¶KEYS.F
C VARIABLES DEFINED TO BREAK DEFAULT TYPES.
      INTEGER CLASS,TYPE
      REAL IBP,IDM,IIP,IIPIP,INP,IOM,IP,IPDM,IUP,JOUR,LPA,LPI,LPN,
     2LPNIP,LPNKG,LPNLPA,LWG,ME,MEDM,MILKKG,MPA,MPN,MPNMPA,
     3NCP,NCPBCP,NEG,NEGAIN,NEGDM,NEGLWG,NEM,NEMDM,NEML,NEMM,NEMY
C
C DATA INPUT OR DEFINITION OF ANIMAL
C
      WRITE(1,10)
   10 FORMAT(1X,'ENTER BODY WEIGHT IN KG AS XXX. ')
      READ(1,11)BW
   11 FORMAT(F5.0)
      WRITE(1,20)
   20 FORMAT(1X,'ENTER WEIGHT GAIN IN KG AS XX.XX ')
      READ(1,21)LWG
   21 FORMAT(F5.2)
      IF(LWG.EQ.0.0)GO TO 22
      WRITE(1,30)
   30 FORMAT(1X,'TYPE OPTIONS FOR NEG REQUIRED FOR GAIN ARE:')
      WRITE(1,31)
   31 FORMAT(1X,'1. MEDIUM-FRAME STEER CALVES')
      WRITE(1,32)
   32 FORMAT(1X,'2. LARGE-FRAME STEER CALVES,COMPENSATING MEDIUM-FRAME')
      WRITE(1,33)
   33 FORMAT(4X,'YEARLING STEERS, AND MEDIUM-FRAME BULL CALVES')
      WRITE(1,34)
   34 FORMAT(1X,'3. LARGE-FRAME BULL CALVES AND COMPENSATING LARGE-FRAME
     2')
      WRITE(1,35)
   35 FORMAT(4X,'YEARLING STEERS')
      WRITE(1,36)
   36 FORMAT(1X,'4. MEDIUM-FRAME HEIFER CALVES')
      WRITE(1,37)
   37 FORMAT(1X,'5. LARGE-FRAME HEIFER CALVES AND COMPENSATING YEARLING
     2HEIFERS')
      WRITE(1,38)
   38 FORMAT(1X,'6. MATURE THIN COWS')
      WRITE(1,39)
   39 FORMAT(1X,'ENTER TYPE AS X')
      READ(1,40)TYPE
   40 FORMAT(I1)
   22 CONTINUE
      WRITE(1,41)
   41 FORMAT(1X,'ENTER CALF BIRTH WEIGHT AS XX. ')
      READ(1,11)CBW
      IF(CBW.EQ.0.0)GO TO 42
      WRITE(1,43)
   43 FORMAT(1X,'ENTER DAYS PREGNANT AS XXX. ')
      READ(1,11)DAYS
   42 CONTINUE
      WRITE(1,50)
   50 FORMAT(1X,'ENTER MILK PRODUCTION IN KG AS XX. ')
      READ(1.11)MILKKG
```

APPENDIX TABLE 10   (*Continued*)

```
            IF(MILKKG.EQ.0.0)GO TO 51
            WRITE(1,52)
        52  FORMAT(1X,'ENTER PERCENT FAT AS X.XX ')
            READ(1,53)PFAT
        53  FORMAT(F5.2)
        51  CONTINUE
            WRITE(1,54)
        54  FORMAT(1X,'CLASS OPTIONS FOR DRY MATTER INTAKE ARE:')
            WRITE(1,55)
        55  FORMAT(1X,'1. MEDIUM-FRAME STEER CALF, LARGE-FRAME HEIFER,')
            WRITE(1,56)
        56  FORMAT(4X,'AND MEDIUM-FRAME BULL')
            WRITE(1,57)
        57  FORMAT(1X,'2. LARGE-FRAME STEER CALF AND MEDIUM-FRAME YEARLING STE
           2ER')
            WRITE(1,58)
        58  FORMAT(1X,'3. LARGE-FRAME BULLS')
            WRITE(1,59)
        59  FORMAT(1X,'4. MEDIUM-FRAME HEIFERS')
            WRITE(1,49)
        49  FORMAT(1X,'5. BREEDING FEMALES')
            WRITE(1,60)
        60  FORMAT(1X,'ENTER CLASS AS X')
            READ(1,40)CLASS
      C
      C DRY MATTER AND ENERGY RELATIONSHIPS
      C
            NEMM=.077*BW**.75                /*NEM FOR MAINTENANCE, MCAL/DAY
            E=2.7182818                      /*BASE OF NATURAL LOGARITHM
      C NEM FOR PREGNANCY, MCAL/DAY
            NEMY=CBW*(.0149-.0000407*DAYS)*E**(.05883*DAYS-.0000804*DAYS**2)/1
           2000
            IF(DAYS.LE.180)NEMY=0.0          /*STARTS PREG. REQ. AT 180 DAYS
            NEML=MILKKG*(.1*PFAT+.35)        /*NEM FOR LACTATION, MCAL/DAY
            NEM=NEMM+NEMY+NEML               /*NEM TOTAL NEEDED, MCAL/DAY
            GO TO(61,62,63,64,65,66),TYPE    /*USES NEG EQ. BASED ON TYPE
        61  NEG=(.0557*BW**.75)*(LWG**1.097) /*NEG FOR TYPE 1, MCAL/DAY
            GO TO 68
        62  NEG=(.0493*BW**.75)*(LWG**1.097) /*NEG FOR TYPE 2, MCAL/DAY
            GO TO 68
        63  NEG=(.0437*BW**.75)*(LWG**1.097) /*NEG FOR TYPE 3, MCAL/DAY
            GO TO 68
        64  NEG=(.0686*BW**.75)*(LWG**1.119) /*NEG FOR TYPE 4, MCAL/DAY
            GO TO 68
        65  NEG=(.0608*BW**.75)*(LWG**1.119) /*NEG FOR TYPE 5, MCAL/DAY
            GO TO 68
        66  NEG=6.2*LWG                      /*NEG FOR TYPE 6, MCAL/DAY
        68  CONTINUE
            IF(LWG.EQ.0.0)GO TO 67           /*PREVENTS DIVISION BY ZERO
            NEGLWG=NEG/LWG                   /*MCAL NEG/KG LIVE WEIGHT GAIN
        67  CONTINUE
      C ITERATIVE LOOP FOR SOLVING ME, MCAL ME/KG DM
            N=0                              /*SETS ITERATION COUNTER TO 0
            MEDM=1.8                         /*SETS INITIAL ME AT 1.8 MCAL/KG DM
        70  CONTINUE
            NEMDM=1.37*MEDM-.138*MEDM**2+.0105*MEDM**3-1.12 /*NEM, MCAL/KG DM
            NEGDM=1.42*MEDM-.174*MEDM**2+.0122*MEDM**3-1.65 /*NEG, MCAL/KG DM
            GO TO(71,72,73,74,75),CLASS  /*USES DM INTAKE EQ. BASED ON CLASS
      C DM INTAKE FOR CLASS 1, KG/DAY
        71  DM=((1.00*BW)**.75)*(.1493*NEMDM-.0460*NEMDM**2-.0196)
            GO TO 76
      C DM INTAKE FOR CLASS 2, KG/DAY
        72  DM=((1.10*BW)**.75)*(.1493*NEMDM-.0460*NEMDM**2-.0196)
            GO TO 76
      C DM INTAKE FOR CLASS 3, KG/DAY
        73  DM=((1.05*BW)**.75)*(.1493*NEMDM-.0460*NEMDM**2-.0196)
```

APPENDIX TABLE 10   (*Continued*)

```
                     GO TO 76
 C DM INTAKE FOR CLASS 4, KG/DAY
       74 DM=((0.90*BW)**.75)*(.1493*NEMDM-.0460*NEMDM**2-.0196)
                     GO TO 76
 C DM INTAKE FOR CLASS 5, KG/DAY
       75 DM=(BW**.75)*(.1462*NEMDM-.0517*NEMDM**2-.0074)
       76 CONTINUE
          DMM=NEM/NEMDM                     /*DM FOR MAINTENANCE, KG
          DMG=NEG/NEGDM                     /*DM FOR GROWTH, KG
          DMT=DMM+DMG                       /*DM TOTAL REQUIRED, KG
          ME=MEDM*DMT                       /* CALCULATES ME, MCAL
          N=N+1                             /*COUNTS ITERATION CYCLE
          IF(N.GT.25)GO TO 78                 /*SETS ITERATION LIMIT
          IF(ABS(DM-DMT).LE. .02)GO TO 78   /*DECISION TO END ITERATION
          IF(DM.LT.DMT)GO TO 79             /*DECISION TO INCREASE MEADM
          IF(DM.GT.DMT)GO TO 80             /*DECISION TO DECREASE MEADM
       79 CONTINUE
          MEDM=MEDM+.1                      /*INCREASE MEDM BY +.1
          GO TO 70
       80 CONTINUE
          MEDM=MEDM-.01                     /*DECREASE MEDM BY -.01
          GO TO 70
 C END OF ITERATIVE LOOP
       78 CONTINUE
          CONDM=(MEDM-2.0)/1.2              /*CONCENTRATE DM INTAKE, PROPORTION DM
          IF(CONDM.LT.0.0)CONDM=0.0         /*PREVENTS NEGATIVE CONCENTRATE INTAKE
          FORDM=1.0-CONDM                   /*FORAGE DM INTAKE, PROPORTION DM
          DMBW=DM/BW                        /*DM INTAKE, PROPORTION BODY WEIGHT
          CI=DMBW*CONDM*100                 /*CONCENTRATE DM INTAKE, PROPORTION BW
          FI=DMBW*FORDM*100                 /*FORAGE DM INTAKE, PROPORTION BW
          DE=ME/.82                         /*DE INTAKE, MCAL/DAY
          TDN=DE/4.4                        /*TDN INTAKE, KG/DAY
          BTDN=TDN                          /*BASELINE TDN (TOTAL), KG
          IDM=DM-BTDN                       /*INDIGESTIBLE DM, KG/DAY
 C
 C STATED PROTEIN FACTORS WITH PROPORTIONAL UNITS
 C
          BTPBCP=.80    /*BACTERIAL TRUE PROTEIN/BACTERIAL CRUDE PROTEIN
          DBPBTP=.80    /*DIGESTIBLE BACTERIAL PROTEIN/BACTERIAL TRUE PROTEIN
          DUPUIP=.80    /*DIGESTIBLE UNDEGRADED PROTEIN/UNDEGRADED INTAKE PROTEIN
          MPNMPA=.67    /*MAINTENANCE PROTEIN NET/MAINTENANCE PROTEIN ABSORBED
          YPNYPA=.50    /*CONCEPTUS PROTEIN NET/CONCEPTUS PROTEIN ABSORBED
          FPAIDM=.090   /*(METABOLIC) FECAL PROTEIN ABSORBED/INDIGESTIBLE DRY MATTER
          DNPNCP=1.00   /*DIGESTIBLE NUCLEIC PROTEIN/NUCLEIC CRUDE PROTEIN
          BCPRAP=.90    /*BACTERIAL CRUDE PROTEIN/RUMEN AVAILABLE PROTEIN
          RIPIP=.15     /*RUMEN INFLUX PROTEIN/INTAKE PROTEIN
          LPNLPA=.65    /*LACTATION PROTEIN NET/LACTATION PROTEIN ABSORBED
          RPNRPA=.50    /*RETAINED PROTEIN NET/RETAINED PROTEIN ABSORBED
 C
 C CALCULATION OF PROTEIN REQUIREMENTS
 C
          SPN= .2 *BW**.6            /*SCURF PROTEIN NET, G
          UPN=2.75*BW**.5            /*ENDOGENOUS URINARY PROTEIN NET, G
          SPA=SPN/MPNMPA             /*SCURF PROTEIN ABSORBED, G
          UPA=UPN/MPNMPA             /*ENDOGENOUS URINARY PROTEIN ABSORBED, G
          MPA=SPA+UPA                /*MAINTENANCE  PROTEIN ABSORBED, G
          FPA=IDM*FPAIDM*1000        /*METABOLIC FECAL PROTEIN ABSORBED, G
          LPN=MILKKG*33.5            /*LACTATION PROTEIN NET, G
          LPA=LPN/LPNLPA             /*LACTATION PROTEIN ABSORBED, G
          YPN=55.                    /*CONCEPTUS PROTEIN NET, G
          IF(DAYS.LE.180)YPN=0.0     /*STARTS PREG. REQ. AT 180 DAYS
          YPA=YPN/YPNYPA             /*CONCEPTUS PROTEIN ABSORBED, G
          RPN=(268-29.4*NEGLWG)*LWG  /*RETAINED PROTEIN NET, G
          RPA=RPN/RPNRPA             /*RETAINED PROTEIN ABSORBED, G
          AP=MPA+LPA+FPA+YPA+RPA     /*ABSORBED PROTEIN (TOTAL), G
 C FLOW OF PROTEIN
```

APPENDIX TABLE 10   (*Continued*)

```
                    C BACTERIAL CRUDE PROTEIN, G
                          BCP=6.25*BTDN*(8.63+14.60*FI-5.18*FI**2+.59*CI)
                          RAP=BCP/BCPRAP                /*RUMEN AVAILABLE PROTEIN, G
                          BTP=BCP*BTPBCP                /*BACTERIAL TRUE PROTEIN, G
                          DBP=BTP*DBPBTP                /*DIGESTIBLE BACTERIAL PROTEIN, G
                          IBP=BTP-DBP                   /*INDIGESTIBLE BACTERIAL PROTEIN, G
                          NCP=BCP-BTP                   /*NUCLEIC CRUDE PROTEIN, G
                          DNP=NCP*DNPNCP                /*DIGESTIBLE NUCLEIC PROTEIN, G
                          INP=NCP-DNP                   /*INDIGESTIBLE NUCLEIC PROTEIN, G
                          DUP=AP-DBP                    /*DIGESTIBLE UNDEGRADED PROTEIN, G
                          UIP=DUP/DUPUIP                /*UNDEGRADED INTAKE PROTEIN, G
                          IUP=UIP-DUP                   /*INDIGESTIBLE UNDEGRADED PROTEIN, G
                          REP=RAP*(1-BCPRAP)            /*RUMEN EFFLUX PROTEIN, G
                          LPI=LPA-LPN                   /*LACTATION PROTEIN INCREMENT, G
                          YPI=YPA-YPN                   /*CONCEPTUS PROTEIN INCREMENT, G
                          RPI=RPA-RPN                   /*RETAINED PROTEIN INCREMENT, G
                          SPI=SPA-SPN                   /*SCURF PROTEIN INCREMENT, G
                          IP=(RAP+UIP)/(1+RIPIP)        /*INTAKE PROTEIN, G
                          IPDM=IP/(DM*1000)             /*INTAKE PROTEIN/DRY MATTER, PROPORTIONAL
                          RIP=RIPIP*IP                  /*RUMEN INFLUX PROTEIN, G
                          DIP=RAP-RIP                   /*DEGRADABLE INTAKE PROTEIN, G
                          UIPIP=UIP/IP                  /*UNDEGRADED INTAKE PROTEIN/INTAKE PROTEIN, G
                    C OUTPUT OF PROTEIN
                          FP=IBP+INP+IUP+FPA            /*FECAL PROTEIN, G
                          UP=REP+DNP+MPA+LPI+YPI+RPI-RIP-SPN /*URINARY PROTEIN, G
                          LPNIP=LPN/IP           /*LACTATION PROTEIN NET/INTAKE PROTEIN, PROPORTIONAL
                          YPNIP=YPN/IP           /*CONCEPTUS PROTEIN NET/INTAKE PROTEIN, PROPORTIONAL
                          RPNIP=RPN/IP           /*RETAINED PROTEIN NET/INTAKE PROTEIN, PROPORTIONAL
                          SPNIP=SPN/IP           /*SCURF PROTEIN NET/INTAKE PROTEIN, PROPORTION
                          UPIP=UP/IP             /*URINARY PROTEIN/INTAKE PROTEIN, PROPORTIONAL
                          FPIP=FP/IP             /*FECAL PROTEIN/INTAKE PROTEIN, PROPORTIONAL
                    C PREDICTED PROTEIN AT DUODENUM
                          STP=BTP+UIP                   /*SMALL (INTESTINE) TRUE PROTEIN, G
                          SCP=BCP+UIP                   /*SMALL (INTESTINE) CRUDE PROTEIN, G
                    C EXPECTED PROTEIN AT DUODENUM
                          TAMM=((32.3*DOM)-8.63)*6.25
                          ROHR=((31.42*DOM)-40.56)*6.25
                          JOUR=((22.62*DOM)+(.687*UIP/6.25)+4.3)*6.25
                          VERI=((23.85*DOM)+(.600*UIP/6.25)+8.6)*6.25
                    C
                    C DATA OUTPUT OR PRINTING OF RESULTS
                    C
                          CALL T10U(:214)                    /*STARTS NEW PAGE
                          WRITE(1,90)
                       90 FORMAT(/)
                          WRITE(1,100)IPDM
                      100 FORMAT('INTAKE PROTEIN/DRY MATTER IS                    ',F5.4,)
                          WRITE(1,110)UIPIP
                      110 FORMAT('UNDEGRADED INTAKE PROTEIN/INTAKE PROTEIN IS      ',F5.3,)
                          WRITE(1,120)BTDN
                      120 FORMAT('BASELINE TDN INTAKE IS                   ',F5.2,' KG')
                          WRITE(1,121)CI
                      121 FORMAT('CONCENTRATE DM INTAKE AS PERCENT BW IS           ',F5.2,)
                          WRITE(1,122)FI
                      122 FORMAT('FORAGE DM INTAKE AS PERCENT OF BW IS             ',F5.2,)
                          WRITE(1,130)DM
                      130 FORMAT('DRY MATTER INTAKE IS                 ',F5.2,' KG')
                          WRITE(1,140)MEDM
                      140 FORMAT('MCAL ME/KG DM IS                             ',F5.2,)
                          WRITE(1,150)NEMDM
                      150 FORMAT('MCAL NEM/KG DM IS                            ',F5.2,)
                          WRITE(1,160)NEGDM
                      160 FORMAT('MCAL NEG/KG DM IS                            ',F5.2,)
                          WRITE(1,170)ME
                      170 FORMAT('MCAL ME NEEDED IS                            ',F5.2,)
                          WRITE(1.180)NEM
```

APPENDIX TABLE 10   (*Continued*)

```
  180 FORMAT('MCAL NEM NEEDED IS                           ',F5.2,)
      WRITE(1,190)NEG
  190 FORMAT('MCAL NEG NEEDED IS                           ',F5.2,)
      WRITE(1,90)
      WRITE(1,200)SPA
  200 FORMAT('SURFACE PROTEIN IN ABSORBED UNITS IS         ',F5.0,' G')
      WRITE(1,210)UPA
  210 FORMAT('URINARY PROTEIN IN ABSORBED UNITS IS         ',F5.0,' G')
      WRITE(1,220)MPA
  220 FORMAT('MAINTENANCE PROTEIN IN ABSORBED UNITS IS     ',F5.0,' G')
      WRITE(1,230)FPA
  230 FORMAT('MET. FECAL PROTEIN IN ABSORBED UNITS IS      ',F5.0,' G')
      WRITE(1,240)LPA
  240 FORMAT('LACTATION PROTEIN IN ABSORBED UNITS IS       ',F5.0,' G')
      WRITE(1,250)YPA
  250 FORMAT('CONCEPTUS PROTEIN IN ABSORBED UNITS IS       ',F5.0,' G')
      WRITE(1,270)RPA
  270 FORMAT('RETAINED PROTEIN IN ABSORBED UNITS IS        ',F5.0,' G')
      WRITE(1,280)AP
  280 FORMAT('REQUIRED PROTEIN IN ABSORBED UNITS IS        ',F5.0,' G')
      WRITE(1,90)
      WRITE(1,300)BCP
  300 FORMAT('BACTERIAL CRUDE PROTEIN IS                   ',F5.0,' G')
      WRITE(1,310)NCP
  310 FORMAT('NUCLEIC CRUDE PROTEIN IS                     ',F5.0,' G')
      WRITE(1,320)BTP
  320 FORMAT('BACTERIAL TRUE PROTEIN IS                    ',F5.0,' G')
      WRITE(1,330)DBP
  330 FORMAT('DIGESTIBLE BACTERIAL PROTEIN IS              ',F5.0,' G')
      WRITE(1,340)DUP
  340 FORMAT('DIGESTIBLE UNDEGRADED PROTEIN IS             ',F5.0,' G')
      WRITE(1,350)UIP
  350 FORMAT('UNDEGRADED INTAKE PROTEIN IS                 ',F5.0,' G')
      WRITE(1,360)DIP
  360 FORMAT('DEGRADED INTAKE PROTEIN IS                   ',F5.0,' G')
      WRITE(1,370)IP
  370 FORMAT('INTAKE PROTEIN IS                            '
     2,F5.0,' G')
      WRITE(1,90)
      WRITE(1,380)SPN
  380 FORMAT('SCURF PROTEIN IS                             '
     2,F5.0,' G')
      WRITE(1,400)RIP
  400 FORMAT('RUMEN INFLUX PROTEIN IS                      ',F5.0,' G')
      WRITE(1,410)REP
  410 FORMAT('RUMEN EFFLUX PROTEIN IS                      ',F5.0,' G')
      WRITE(1,420)DNP
  420 FORMAT('DIGESTIBLE NUCLEIC PROTEIN IS                ',F5.0,' G')
      WRITE(1,430)MPA
  430 FORMAT('MAINTENANCE PROTEIN IN ABSORBED UNITS IS     ',F5.0,' G')
      WRITE(1,440)LPI
  440 FORMAT('LACTATION PROTEIN INCREMENT IS               ',F5.0,' G')
      WRITE(1,450)YPI
  450 FORMAT('PREGNANCY PROTEIN INCREMENT IS               ',F5.0,' G')
      WRITE(1,460)RPI
  460 FORMAT('RETAINED PROTEIN INCREMENT IS                ',F5.0,' G')
      WRITE(1,470)UP
  470 FORMAT('URINARY PROTEIN IS                           '
     2,F5.0,' G')
      WRITE(1,90)
      WRITE(1,500)IBP
  500 FORMAT('INDIGESTIBLE BACTERIAL PROTEIN IS            ',F5.0,' G')
      WRITE(1,510)INP
  510 FORMAT('INDIGESTIBLE NUCLEIC PROTEIN IS              ',F5.0,' G')
      WRITE(1,520)IUP
  520 FORMAT('INDIGESTIBLE UNDEGRADED PROTEIN IS           ',F5.0,' G')
```

APPENDIX TABLE 10   *(Continued)*

```
      WRITE(1,530)FPA
  530 FORMAT('METABOLIC FECAL PROTEIN IN ABSORBED UNITS IS  ',F5.0,' G')
      WRITE(1,540)FP
  540 FORMAT('FECAL PROTEIN IS                              '
     2,F5.0,' G')
      WRITE(1,90)
      WRITE(1,600)LPN
  600 FORMAT('MILK PROTEIN IN NET UNITS IS                  '
     2,F5.0,' G')
      WRITE(1,610)YPN
  610 FORMAT('CONCEPTUS PROTEIN IN NET UNITS IS             '
     2,F5.0,' G')
      WRITE(1,620)RPN
  620 FORMAT('RETAINED PROTEIN IN NET UNITS IS              '
     2,F5.0,' G')
      WRITE(1,90)
      WRITE(1,700)LPNIP
  700 FORMAT('MILK PROTEIN IN NET UNITS/INTAKE PROTEIN IS    ',F5.3,)
      WRITE(1,710)YPNIP
  710 FORMAT('CONCEPTUS PROTEIN NET/INTAKE PROTEIN IS        ',F5.3,)
      WRITE(1,720)DPAIP
  720 FORMAT('DIFFERENCE PROTEIN ABSORBED/INTAKE PROTEIN IS  ',F5.3,)
      WRITE(1,730)RPNIP
  730 FORMAT('RETAINED PROTEIN NET/INTAKE PROTEIN IS         ',F5.3,)
      WRITE(1,740)SPNIP
  740 FORMAT('SCURF PROTEIN NET/INTAKE PROTEIN IS            ',F5.3,)
      WRITE(1,750)UPIP
  750 FORMAT('URINARY PROTEIN/INTAKE PROTEIN IS              ',F5.3,)
      WRITE(1,760)FPIP
  760 FORMAT('FECAL PROTEIN/INTAKE PROTEIN IS                ',F5.3,)
      CALL EXIT
      END
OK,
```

APPENDIX TABLE 10   (*Continued*)

```
      R NRCB84
   ENTER BODY WEIGHT IN KG AS XXX.
300.
   ENTER WEIGHT GAIN IN KG AS XX.XX
1.2
   TYPE OPTIONS FOR NEG REQUIRED FOR GAIN ARE:
   1. MEDIUM-FRAME STEER CALVES
   2. LARGE-FRAME STEER CALVES,COMPENSATING MEDIUM-FRAME
      YEARLING STEERS, AND MEDIUM-FRAME BULL CALVES
   3. LARGE-FRAME BULL CALVES AND COMPENSATING LARGE-FRAME
      YEARLING STEERS
   4. MEDIUM-FRAME HEIFER CALVES
   5. LARGE-FRAME HEIFER CALVES AND COMPENSATING YEARLING HEIFERS
   6. MATURE THIN COWS
   ENTER TYPE AS X
1
   ENTER CALF BIRTH WEIGHT AS XX.
0.
   ENTER MILK PRODUCTION IN KG AS XX.
0.
   CLASS OPTIONS FOR DRY MATTER INTAKE ARE:
   1. MEDIUM-FRAME STEER CALF, LARGE-FRAME HEIFER,
      AND MEDIUM-FRAME BULL
   2. LARGE-FRAME STEER CALF AND MEDIUM-FRAME YEARLING STEER
   3. LARGE-FRAME BULLS
   4. MEDIUM-FRAME HEIFERS
   5. BREEDING FEMALES
   ENTER CLASS AS X
1


   INTAKE PROTEIN/DRY MATTER IS                      .1131
   UNDEGRADED INTAKE PROTEIN/INTAKE PROTEIN IS      0.285
   BASELINE TDN INTAKE IS                      5.47 KG
   CONCENTRATE DM INTAKE AS PERCENT BW IS           1.48
   FORAGE DM INTAKE AS PERCENT OF BW IS             0.92
   DRY MATTER INTAKE IS                        7.20 KG
   MCAL ME/KG DM IS                                 2.74
   MCAL NEM/KG DM IS                                1.81
   MCAL NEG/KG DM IS                                1.19
   MCAL ME NEEDED IS                               19.72
   MCAL NEM NEEDED IS                               5.55
   MCAL NEG NEEDED IS                               4.90

   SURFACE PROTEIN IN ABSORBED UNITS IS             9. G
   URINARY PROTEIN IN ABSORBED UNITS IS            71. G
   MAINTENANCE PROTEIN IN ABSORBED UNITS IS        80. G
   MET. FECAL PROTEIN IN ABSORBED UNITS IS        156. G
   LACTATION PROTEIN IN ABSORBED UNITS IS           0. G
   CONCEPTUS PROTEIN IN ABSORBED UNITS IS           0. G
   RETAINED PROTEIN IN ABSORBED UNITS IS          355. G
   REQUIRED PROTEIN IN ABSORBED UNITS IS          591. G

   BACTERIAL CRUDE PROTEIN IS                     634. G
   NUCLEIC CRUDE PROTEIN IS                       127. G
   BACTERIAL TRUE PROTEIN IS                      507. G
   DIGESTIBLE BACTERIAL PROTEIN IS                406. G
   DIGESTIBLE UNDEGRADED PROTEIN IS               186. G
   UNDEGRADED INTAKE PROTEIN IS                   232. G
   DEGRADED INTAKE PROTEIN IS                     582. G
   INTAKE PROTEIN IS                                      814. G

   SCURF PROTEIN IS                                        6. G
   RUMEN INFLUX PROTEIN IS                        122. G
   RUMEN EFFLUX PROTEIN IS                         70. G
   DIGESTIBLE NUCLEIC PROTEIN IS                  127. G
```

APPENDIX TABLE 10   (*Continued*)

```
MAINTENANCE PROTEIN IN ABSORBED UNITS IS         80. G
LACTATION PROTEIN INCREMENT IS                    0. G
PREGNANCY PROTEIN INCREMENT IS                    0. G
RETAINED PROTEIN INCREMENT IS                   177. G
URINARY PROTEIN IS                                            327. G

INDIGESTIBLE BACTERIAL PROTEIN IS               101. G
INDIGESTIBLE NUCLEIC PROTEIN IS                   0. G
INDIGESTIBLE UNDEGRADED PROTEIN IS               46. G
METABOLIC FECAL PROTEIN IN ABSORBED UNITS IS    156. G
FECAL PROTEIN IS                                             304. G

MILK PROTEIN IN NET UNITS IS                                   0. G
CONCEPTUS PROTEIN IN NET UNITS IS                             0. G
RETAINED PROTEIN IN NET UNITS IS                            177. G

MILK PROTEIN IN NET UNITS/INTAKE PROTEIN IS     0.000
CONCEPTUS PROTEIN NET/INTAKE PROTEIN IS         0.000
DIFFERENCE PROTEIN ABSORBED/INTAKE PROTEIN IS   0.000
RETAINED PROTEIN NET/INTAKE PROTEIN IS          0.218
SCURF PROTEIN NET/INTAKE PROTEIN IS             0.008
URINARY PROTEIN/INTAKE PROTEIN IS               0.401
FECAL PROTEIN/INTAKE PROTEIN IS                 0.373
OK,
```

APPENDIX TABLE 10   (*Continued*)

```
      R NRCB84
   ENTER BODY WEIGHT IN KG AS XXX.
500.
   ENTER WEIGHT GAIN IN KG AS XX.XX
0.
   ENTER CALF BIRTH WEIGHT AS XX.
36.
   ENTER DAYS PREGNANT AS XXX.
225.
   ENTER MILK PRODUCTION IN KG AS XX.
0.
   CLASS OPTIONS FOR DRY MATTER INTAKE ARE:
   1. MEDIUM-FRAME STEER CALF, LARGE-FRAME HEIFER,
      AND MEDIUM-FRAME BULL
   2. LARGE-FRAME STEER CALF AND MEDIUM-FRAME YEARLING STEER
   3. LARGE-FRAME BULLS
   4. MEDIUM-FRAME HEIFERS
   5. BREEDING FEMALES
   ENTER CLASS AS X
5

   INTAKE PROTEIN/DRY MATTER IS                      .0864
   UNDEGRADED INTAKE PROTEIN/INTAKE PROTEIN IS      0.397
   BASELINE TDN INTAKE IS                          5.02 KG
   CONCENTRATE DM INTAKE AS PERCENT BW IS           0.00
   FORAGE DM INTAKE AS PERCENT OF BW IS             1.90
   DRY MATTER INTAKE IS                            9.49 KG
   MCAL ME/KG DM IS                                 1.91
   MCAL NEM/KG DM IS                                1.07
   MCAL NEG/KG DM IS                                0.51
   MCAL ME NEEDED IS                               18.13
   MCAL NEM NEEDED IS                              10.12
   MCAL NEG NEEDED IS                               0.00

   SURFACE PROTEIN IN ABSORBED UNITS IS             12. G
   URINARY PROTEIN IN ABSORBED UNITS IS             92. G
   MAINTENANCE PROTEIN IN ABSORBED UNITS IS        104. G
   MET. FECAL PROTEIN IN ABSORBED UNITS IS         402. G
   LACTATION PROTEIN IN ABSORBED UNITS IS            0. G
   CONCEPTUS PROTEIN IN ABSORBED UNITS IS          110. G
   RETAINED PROTEIN IN ABSORBED UNITS IS             0. G
   REQUIRED PROTEIN IN ABSORBED UNITS IS           616. G

   BACTERIAL CRUDE PROTEIN IS                      555. G
   NUCLEIC CRUDE PROTEIN IS                         111. G
   BACTERIAL TRUE PROTEIN IS                       444. G
   DIGESTIBLE BACTERIAL PROTEIN IS                 355. G
   DIGESTIBLE UNDEGRADED PROTEIN IS                260. G
   UNDEGRADED INTAKE PROTEIN IS                    326. G
   DEGRADED INTAKE PROTEIN IS                      494. G
   INTAKE PROTEIN IS                                       820. G

   SCURF PROTEIN IS                                         8. G
   RUMEN INFLUX PROTEIN IS                         123. G
   RUMEN EFFLUX PROTEIN IS                          62. G
   DIGESTIBLE NUCLEIC PROTEIN IS                   111. G
   MAINTENANCE PROTEIN IN ABSORBED UNITS IS        104. G
   LACTATION PROTEIN INCREMENT IS                    0. G
   PREGNANCY PROTEIN INCREMENT IS                   55. G
   RETAINED PROTEIN INCREMENT IS                     0. G
   URINARY PROTEIN IS                                     201. G

   INDIGESTIBLE BACTERIAL PROTEIN IS                89. G
   INDIGESTIBLE NUCLEIC PROTEIN IS                   0. G
   INDIGESTIBLE UNDEGRADED PROTEIN IS               65. G
```

APPENDIX TABLE 10   (*Continued*)

```
          METABOLIC FECAL PROTEIN IN ABSORBED UNITS IS    402. G
          FECAL PROTEIN IS                                          556. G

          MILK PROTEIN IN NET UNITS IS                               0. G
          CONCEPTUS PROTEIN IN NET UNITS IS                         55. G
          RETAINED PROTEIN IN NET UNITS IS                           0. G

          MILK PROTEIN IN NET UNITS/INTAKE PROTEIN IS    0.000
          CONCEPTUS PROTEIN NET/INTAKE PROTEIN IS        0.067
          DIFFERENCE PROTEIN ABSORBED/INTAKE PROTEIN IS  0.000
          RETAINED PROTEIN NET/INTAKE PROTEIN IS         0.000
          SCURF PROTEIN NET/INTAKE PROTEIN IS            0.010
          URINARY PROTEIN/INTAKE PROTEIN IS              0.245
          FECAL PROTEIN/INTAKE PROTEIN IS                0.678
          OK,
```

APPENDIX TABLE 10   (*Continued*)

```
     R NRCB84
  ENTER BODY WEIGHT IN KG AS XXX.
500.
  ENTER WEIGHT GAIN IN KG AS XX.XX
0.
  ENTER CALF BIRTH WEIGHT AS XX.
0.
  ENTER MILK PRODUCTION IN KG AS XX.
10.
  ENTER PERCENT FAT AS X.XX
4.00
  CLASS OPTIONS FOR DRY MATTER INTAKE ARE:
  1. MEDIUM-FRAME STEER CALF, LARGE-FRAME HEIFER,
     AND MEDIUM-FRAME BULL
  2. LARGE-FRAME STEER CALF AND MEDIUM-FRAME YEARLING STEER
  3. LARGE-FRAME BULLS
  4. MEDIUM-FRAME HEIFERS
  5. BREEDING FEMALES
  ENTER CLASS AS X
5

INTAKE PROTEIN/DRY MATTER IS                        .1207
UNDEGRADED INTAKE PROTEIN/INTAKE PROTEIN IS         0.400
BASELINE TDN INTAKE IS                          6.80 KG
CONCENTRATE DM INTAKE AS PERCENT BW IS             0.75
FORAGE DM INTAKE AS PERCENT OF BW IS               1.25
DRY MATTER INTAKE IS                           10.03 KG
MCAL ME/KG DM IS                                   2.45
MCAL NEM/KG DM IS                                  1.56
MCAL NEG/KG DM IS                                  0.96
MCAL ME NEEDED IS                                 24.53
MCAL NEM NEEDED IS                                15.64
MCAL NEG NEEDED IS                                 0.00

SURFACE PROTEIN IN ABSORBED UNITS IS               12. G
URINARY PROTEIN IN ABSORBED UNITS IS               92. G
MAINTENANCE PROTEIN IN ABSORBED UNITS IS          104. G
MET. FECAL PROTEIN IN ABSORBED UNITS IS           291. G
LACTATION PROTEIN IN ABSORBED UNITS IS            515. G
CONCEPTUS PROTEIN IN ABSORBED UNITS IS              0. G
RETAINED PROTEIN IN ABSORBED UNITS IS               0. G
REQUIRED PROTEIN IN ABSORBED UNITS IS             910. G

BACTERIAL CRUDE PROTEIN IS                        817. G
NUCLEIC CRUDE PROTEIN IS                          163. G
BACTERIAL TRUE PROTEIN IS                         654. G
DIGESTIBLE BACTERIAL PROTEIN IS                   523. G
DIGESTIBLE UNDEGRADED PROTEIN IS                  387. G
UNDEGRADED INTAKE PROTEIN IS                      484. G
DEGRADED INTAKE PROTEIN IS                        726. G
INTAKE PROTEIN IS                                        1210. G

SCURF PROTEIN IS                                           8. G
RUMEN INFLUX PROTEIN IS                           182. G
RUMEN EFFLUX PROTEIN IS                            91. G
DIGESTIBLE NUCLEIC PROTEIN IS                     163. G
MAINTENANCE PROTEIN IN ABSORBED UNITS IS          104. G
LACTATION PROTEIN INCREMENT IS                    180. G
PREGNANCY PROTEIN INCREMENT IS                      0. G
RETAINED PROTEIN INCREMENT IS                       0. G
URINARY PROTEIN IS                                       349. G

INDIGESTIBLE BACTERIAL PROTEIN IS                 131. G
INDIGESTIBLE NUCLEIC PROTEIN IS                     0. G
INDIGESTIBLE UNDEGRADED PROTEIN IS                 97. G
```

APPENDIX TABLE 10   *(Continued)*

```
            METABOLIC FECAL PROTEIN IN ABSORBED UNITS IS   291. G
            FECAL PROTEIN IS                                         518. G

            MILK PROTEIN IN NET UNITS IS                             335. G
            CONCEPTUS PROTEIN IN NET UNITS IS                          0. G
            RETAINED PROTEIN IN NET UNITS IS                           0. G

            MILK PROTEIN IN NET UNITS/INTAKE PROTEIN IS    0.277
            CONCEPTUS PROTEIN NET/INTAKE PROTEIN IS        0.000
            DIFFERENCE PROTEIN ABSORBED/INTAKE PROTEIN IS  0.000
            RETAINED PROTEIN NET/INTAKE PROTEIN IS         0.000
            SCURF PROTEIN NET/INTAKE PROTEIN IS            0.007
            URINARY PROTEIN/INTAKE PROTEIN IS              0.288
            FECAL PROTEIN/INTAKE PROTEIN IS                0.428
            OK,
```

APPENDIX TABLE 11   Apparent Absorption of Nitrogen from the Small Intestine of Lactating Cattle

DIET	NAN	Amino Acids	Reference
Hay or dried grass			
rolled barley		0.66	ARC (1980)
high moisture barley		0.77	ARC (1980)
pelleted barley		0.69	ARC (1980)
pelleted corn		0.70	ARC (1980)
Alfalfa			
Ensiled at 29% DM	0.62	0.71	Merchen and Satter (1983b)
40% DM	0.61	0.71	Merchen and Satter (1983b)
66% DM	0.63	0.74	Merchen and Satter (1983b)
Hay	0.60	0.73	Merchen and Satter (1983b)
Alfalfa			
Ensiled at 47% DM		0.68	Merchen (1981)
Ensiled at 61% DM		0.66	Merchen (1981)
Corn silage, hay, grain			
with 50Δ protein from			
Soybean meal	0.67	0.71	Stern et al. (1984)
Raw soybeans	0.64	0.70	Stern et al. (1984)
Extruded soybeans 270F	0.70	0.76	Stern et al. (1984)
Extruded soybeans 300F	0.72	0.75	Stern et al. (1984)
Soybean meal	0.63	0.70	Stern and Satter (1982)
Corn gluten meal	0.71	0.76	Stern and Satter (1982)
Wet brewers grains	0.65	0.71	Stern and Satter (1982)
Dried brewers grains			
with solubles	0.58	0.66	Stern and Satter (1982)
Raw whole cottonseeds		0.56	Pena et al. (1985)
Extruded whole cottonseeds		0.60	Pena et al. (1985)
Heated whole cottonseeds		0.63	Pena et al. (1985)

APPENDIX TABLE 12   Apparent Absorption of Nitrogen from the Small Intestine of Nonlactating Cattle

DIET	NAN	Amino Acids	Reference
Hay or dried grass			
rolled barley		0.53	McMeniman (1976); ARC (1980)
rolled barley		0.62	McMeniman (1976); ARC (1980)
flaked barley		0.50	McMeniman (1976); ARC (1980)
rolled barley + urea		0.65	McMeniman (1976); ARC (1980)
flaked barley + urea		0.67	McMeniman (1976); ARC (1980)
rolled barley + unheated beans		0.58	McMeniman (1976); ARC (1980)
rolled barley + heated beans		0.59	McMeniman (1976); ARC (1980)
1 (as designed in	0.64		Zinn and Owens (1982)
2 publication)	0.67		Zinn and Owens (1982)
3	0.60		Zinn and Owens (1982)
4	0.64		Zinn and Owens (1982)
5	0.66		Zinn and Owens (1982)
6	0.70		Zinn and Owens (1982)
7	0.71		Zinn and Owens (1982)
8	0.73		Zinn and Owens (1982)
9	0.68		Zinn and Owens (1982)
10	0.69		Zinn and Owens (1982)
11	0.69		Zinn and Owens (1982)
12	0.61		Zinn and Owens (1982)
13	0.68		Zinn and Owens (1982)
Purified diets			
rapeseed meal		0.66	Sharma et al. (1974); ARC (1980)
HCHO-treated rapeseed meal		0.65	Sharma et al. (1974); ARC (1980)
Casein		0.65	Sharma et al. (1974); ARC (1980)
HCHO-treated casein		0.72	Sharma et al. (1974); ARC (1980)

APPENDIX TABLE 13 Apparent Absorption of Nitrogen from the Small Intestine of Sheep

DIET	NAN	Amino Acids	Reference
Ryegrass		0.71	MacRae and Ulyatt (1974);
		0.74	ARC (1980)
White clover		0.66	Beever et al. (1971); ARC (1980)
Red clover		0.71	
Dried grass	0.64		MacRae et al. (1972); ARC (1980)
Dried grass			
Early cut		0.77	Coelho da Silva et al. (1972a,b);
			ARC (1980)
Medium cut		0.70	
Ground and pelleted grass			
Early cut		0.71	
Medium cut		0.73	
Alfalfa hay	0.60	0.60	
Alfalfa silage, 47% DM	0.54	0.58	Merchen and Satter (1983a)
Alfalfa protein concentrate			
heat coagulated 80C, fresh	0.69		Lu et al. (1981)
heat coagulated 80C, spray dried	0.70		Lu et al. (1981)
fermented, spray dried	0.67		
heat, coagulated 100C, spray dried	0.70		
Hay or dried grass +			
Soybean meal	0.65		MacRae et al. (1972); ARC (1980)
Casein	0.62		
Concentrate	0.48		Ben-Ghedalia et al. (1974);
			ARC (1980)
Dried grass: sugar beet pulp			
1.0:0		0.68	Tas et al. (1981)
0.75:0.25		0.69	Tas et al. (1981)
0.50:0.50		0.71	
0.25:0.75		0.68	
All concentrate			
rolled barley	0.61		Ørskov et al. (1974)
rolled barley	0.65		Ørskov et al. (1972)
rolled barley	0.60		Ørskov et al. (1971a)
rolled barley			
+ urea (0.007)	0.68		Ørskov et al. (1972)
+ urea (0.014)	0.64		Ørskov et al. (1972)
+ urea (0.021)	0.68		Ørskov et al. (1972)
All concentrate (cont'd.)			
rolled barley			
+ urea (0.009)	0.59		Ørskov et al. (1971b)
+ urea (0.020)	0.52		Ørskov et al. (1971b)
+ urea (0.032)	0.59		Ørskov et al. (1971b)
+ urea (0.043)	0.61		Ørskov et al. (1971b)
rolled barley			
+ fish meal (0.038)	0.63		Ørskov et al. (1971b)
+ fish meal (0.089)	0.63		Ørskov et al. (1971b)
+ fish meal (0.141)	0.67		Ørskov et al. (1971b)
+ fish meal (0.192)	0.70		Ørskov et al. (1971b)
+ fish meal (48 g)	0.67		Ørskov et al. (1974)
+ fish meal (94 g)	0.67		Ørskov et al. (1974)
rolled barley			
+ fish meal (48 g), urea (10 g)	0.68		Ørskov et al. (1974)
+ fish meal (48 g), urea (20 g)	0.64		Ørskov et al. (1974)
Ruminal infusion of VFA + minerals, abomasal infusion of microbial isolate + vitamins	0.78		Strom and Ørskov (1979)
Semipurified			
+ 3.6% urea (propionate fermentation)		0.79	Harrison et al. (1976); ARC (1980)
+ 3.6% urea (acetate fermentation)		0.76	Harrison et al. (1976); ARC (1980)
field beans		0.73	McMeniman (1976; ARC (1980)
heated field beans		0.79	McMeniman (1976; ARC (1980)
+ 4.2% urea			
2x feeding		0.65	ARC (1980)
24X feeding		0.70	ARC (1980)
+ lipid		0.68	ARC (1980)
+ VFA		0.56	ARC (1980)

Index